Plant Resilience to Abiotic Stress

Edited by Muhammad Waseem

Published in London, United Kingdom

Plant Resilience to Abiotic Stress
http://dx.doi.org/10.5772/intechopen.1008096
Edited by Muhammad Waseem

Contributors
Abu Bakar Siddique, Aline Chaves Intorne, Amina Ameer, Analyse Villanueva Gaete, Anjum Shaheen, Ansa Asghar, Asima Asghar, Ayesha Siddiqua, Bruna Sthefanie Paz de Souza, Daiane Carvalho Baía, Gabriela Petroceli Mota, Glacielen Ribeiro de Souza, Ilva Licaj, Jazab Shafqat, João Vitor Pereira de Souza, Kennedy Samuel Banja, Mansoor Hameed, Mariapina Rocco, Marília Berbert Molina, Mohammad Shah Jahan, Mst. Mahzabin Zannat, Muhammad Waseem, Nida Hussain, Olusola Jacob Oyetunji, Pingwu Liu, Rizwana Kanwal, Sana Basharat, Sana Saleem, Tahreem Ahmed, Usman Tufail, Zahida Parveen

First published in London, United Kingdom, 2026 by IntechOpen
IntechOpen is the global imprint of INTECHOPEN LIMITED, registered in England and Wales, registration number: 11086078, 167-169 Great Portland Street, London, W1W 5PF, United Kingdom

For EU product safety concerns: IN TECH d.o.o., Prolaz Marije Krucifikse Kozulić 3, 51000 Rijeka, Croatia, info@intechopen.com or visit our website at intechopen.com.

British Library Cataloguing-in-Publication Data
A catalogue record for this book is available from the British Library

Plant Resilience to Abiotic Stress
Edited by Muhammad Waseem
p. cm.
Print ISBN 978-1-83635-529-8
Online ISBN 978-1-83635-528-1
eBook (PDF) ISBN 978-1-83635-530-4

Meet the editor

Dr. Muhammad Waseem is an Associate Professor/Researcher at the School of Breeding and Multiplication (Sanya Institute of Breeding and Multiplication), School of Tropical Agriculture and Forestry, Hainan University, Sanya 572025, China. He received a Ph.D. (specialization in Plant Molecular Biology) from the School of Life Sciences, Chongqing University, Chongqing, PR China. He has 6 years of research experience, including as a postdoctoral researcher. His interests include transcriptional regulation of flowering time, fruit ripening and development, and abiotic stresses such as salinity and drought. In addition, he has hands-on experience in analyzing next-generation sequencing data, transcriptomics, and data visualization. Currently, Dr. Waseem is investigating flowering time regulation in Brassica crops. He was awarded a National Science Foundation Project, RFIS-I in 2023. In brief, he has published more than 62 research articles, 5 review articles, and 9 book chapters in peer-reviewed journals such as *New Phytologist, International Journal of Macromolecules, Planta, GM Crops & Food, Environmental and Experimental Botany, Physiologia Plantarum*, and *Frontiers in Plant Science*.

Contents

Preface

Against the backdrop of intensified global climate change and ongoing fluctuations in the agricultural ecosystem, abiotic stresses such as drought, salt stress, high and low temperatures, and nutrient imbalances are affecting crop productivity and ecosystem stability with unprecedented frequency and intensity. Improving the adaptability and resilience of plants to abiotic stress is not only related to food security and sustainable agricultural development, but also a key scientific proposition for addressing future environmental uncertainty. The book "Plant Resilience to Abiotic Stress" aims to systematically present the response patterns and resilience-enhancement pathways of plants under adversity, from mechanism analysis to application strategies, providing a theoretical framework and research ideas for researchers, graduate students, and agricultural practitioners.

This book first provides an overview of the concept of plant resilience, the main types of abiotic stress, and their impact on growth, development, and yield formation through the introduction chapter "Introducing Plant Resilience to Abiotic Stress", helping readers establish a comprehensive understanding of "stress response adaptation recovery". On this basis, chapter entitled "Molecular and Physiological Responses of Plants to Salinity and Drought Stress: Implications for Crop Improvement and Climate Resilience", focuses on the molecular and physiological responses of plants to salt stress and drought stress, and summarizes current research progress around core processes such as osmotic regulation, ion homeostasis, antioxidant defense, signal transduction, and gene expression regulation, emphasizing the coupling relationship between different hierarchical mechanisms.

Chapter 3 "Phytohormonal Regulation of Photosynthesis under Optimal and Abiotic Stress Conditions" further extends the research perspective to the plant hormone network, discussing how hormones such as ABA, IAA, SA, JA, and ETH participate in stomatal regulation, photosystem stability, carbon assimilation, and energy allocation, thereby achieving a dynamic balance between maintaining photosynthetic efficiency and reducing photoinhibition risk. Chapter 4 "Salt Stress in Crops: Tolerance Mechanisms and Emerging Amelioration Strategies" is guided by crop production, combining the formation mechanism of salt tolerance traits with breeding/management strategies, summarizing emerging directions such as molecular breeding, exogenous regulators, microbial interactions, and comprehensive agronomic measures, aiming to promote the transformation of basic understanding into feasible solutions.

This book, which emphasizes green agriculture and eco-friendly technologies, highlights the potential of biosurfactants to improve the rhizosphere environment, promote nutrient absorption, alleviate oxidative stress, and enhance stress response in Chapter 5, "Abiotic Stress in Plants: Biosurfactants as Tools to Enhance Plant Resilience". Chapter 6 "Serendipita indica: A Plant Elixir for Growth Enhancement and Environmental Stress Resilience" focuses on the endophytic fungus *Serendipita indica* (Cryptococcus/endophytic symbiotic fungus), systematically introducing its mechanism of action and application prospects in promoting growth, improving nutrient utilization efficiency, and enhancing adaptability to various environmental stresses, highlighting the important trend of "plant microbe synergistic resilience".

The book concludes with the perspective chapter "Unlocking Root Architecture Plasticity – Enhancing Crop Productivity and Resilience to Abiotic Stresses", emphasizing the plasticity of root structure as a key hub connecting resource acquisition and adversity adaptation. By analyzing the interactions among root morphology, rhizosphere signals, and environmental feedback, this chapter proposes that improving root resilience will be a breakthrough in balancing yield and stress resistance, providing new research directions and strategic frameworks for future crop improvement and precision agriculture.

We hope this book provides readers with a cross-scale, interdisciplinary perspective, emphasizing both the depth of molecular and physiological mechanisms and the feasibility of agricultural applications and technological transformation. Understanding and enhancing plant resilience under constantly changing environmental conditions will be an important foundation for ensuring stable, high crop yields and promoting agricultural sustainability and ecological security. I dedicate this book to all colleagues and learners who are committed to the study of plant stress biology and to improving crop stress resistance. I hope it becomes a valuable reference for exchanging ideas, inspiring research, and promoting practice.

Muhammad Waseem
School of Breeding and Multiplication (Sanya Institute of Breeding and Multiplication),
School of Tropical Agriculture and Forestry,
Hainan University,
Sanya, China

Chapter 1

Introductory Chapter: Introducing Plant Resilience to Abiotic Stress

Muhammad Waseem

1. Introduction

Plants are exposed to diverse environmental constraints such as salinity, drought, temperature extremes and susceptibility to pathogen invasion throughout their life that adversely impact their growth, development, and final yield and productivity [1, 2]. In the era of climate change, the frequency and magnitude of these stresses become more intense and challenging global food security. To improve crop resilience to abiotic and biotic stresses, an in-depth understanding of plant stress response and tolerance mechanism is required (**Figure 1**). To date, the majority of the studies in this area are primarily based on the experimental studies on plant exposed to solo stress, which rarely occur in field; however, the studies lacking combine stresses. For instance, in field plants exposed to drought and heat often co-occur but flooding and drought occur frequently [3], which is more devastating compared to solo stress [4].

Over the last decade, substantial increase in researches on abiotic stress resilience has been seen. For example, these studies demonstrated the importance of abiotic stresses governed by highly coordinated and intricated networks. This chapter provides an introductory overview of our current understanding of plant response to abiotic tress and recent findings, which enhances our understanding of plant stress resilience to these stresses.

The agriculture industry faces constant environmental stresses throughout its life. Environmental stress negatively affects plant growth and production. Among the environmental stresses, abiotic stress is a significant cause of crop yield loss [5, 6]. All environmental stresses, such as drought, salinity, heavy metals, heat, etc., negatively affect crop production globally. In response to stress, signal transduction pathways become activated [7]. The signal transduction pathway is the regulatory network through which the stress stimulus is sensed, interpreted, and converted into a targeted molecular response [8, 9]. Signal transduction pathways are composed of four main components: signal perception, signal transduction cascade, transcriptional regulation, and ultimately cellular and physiological response. It works as an endogenous communication network that perceives external stimuli and processes them into a specific cellular response [10, 11].

It is a highly coordinated signaling mechanism that adopts the stress response. The signal is transduced through a cascade of molecular events such as secondary messengers (calcium signaling and reactive oxygen species), protein phosphorylation by kinases, and activation of the transcription factors, which regulate gene

IntechOpen

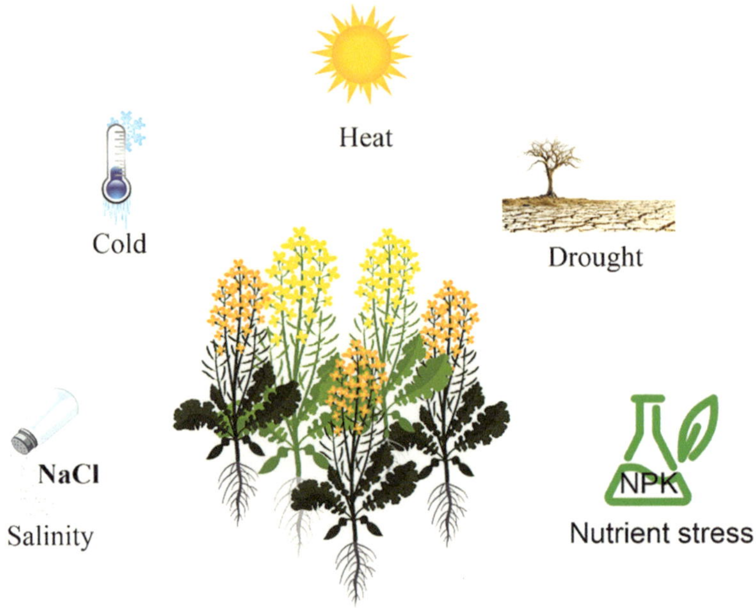

Figure 1.
Diverse abiotic stress on plants such as salinity, drought, temperature extremes (cold and heat), and nutrient stress.

expression [12, 13]. This complex regulatory network ensures the stress tolerance mechanism by the modulatory effects of cellular and physiological responses, such as closure of stomata, metabolic and biochemical reprogramming, activation of phytohormones [14].

2. Perception of stress

In response to the environmental stress, stress perception is the first step that activates the signal transduction pathways. The perception of stress is sensed by receptors such as membrane-bound receptors (RLKs, PRPs) and redox sensors [15–17]. After sensing the stress, calcium ions, reactive oxygen species (ROS), nitric oxide (NO), and secondary messengers such as cyclic nucleotides, IP3, are activated [18, 19]. Calcium signaling is important in abiotic stress responses. Environmental stresses, such as drought, salt, and climate change, induce modulatory changes in calcium concentrations within plant cells. These adaptations are sensed by Ca-binding proteins that respond to stress, introducing a cascade of events in stress mitigation [20]. For example, during drought stress, calcium signaling can facilitate physiological responses such as closure of the stomata to prevent further water loss. In response to salinity, calcium signaling can regulate ion channels to maintain ion homeostasis. Under cold stress, calcium signaling may activate genes that lead to the accumulation of cryoprotectants, which protects plant cells from freezing damage. These responses are complex and involve a network of interactions between signaling pathways. This pathway relates to other signaling molecules such as ROS and phytohormones like abscisic acid (ABA) and mitogen-activated protein kinases (MAPKs), which further modulate the plant's adaptability to stress [8, 21, 22].

Furthermore, ROS production, because of calcium signaling, can activate MAPK cascades, which comprises a series of protein kinases that phosphorylate specific target proteins to elicit a response. This cascade event results in the activation of defense signals and their relative expression of genes [23, 24]. Calcium channels and pumps located on the plasma membrane and endomembrane systems are responsible for the changes in cytosolic Ca^{2+} levels. These changes are essential for activating various defense responses, including ROS production and MAPK activation. In response to stress, there is rapid production of ROS that acts as an endogenous defense mechanism to enhance stress tolerance. In response to ROS, there is oxidative stress in plants that tries to affect several physicochemical and molecular responses in plants. At the same time, ROS acts as a signaling molecule that activates other protective pathways [25, 26].

In response to stress, there is an increase in the level of abscisic acid (ABA). It plays an important role in the signal transduction pathway in stress tolerance by acting as a stress sensor as well as a signal mediator [27, 28]. It immediately activates the antioxidant defense system and tries to regulate the ROS production. Moreover, in stress duration, its role is in the closure of stomata, inhibiting germination, and promoting root elongation under the stress conditions [29]. In stress conditions, ABA interacts with salicylic acid (SA) and jasmonic acid (JA) to regulate the stress response. Moreover, it has an antagonistic effect on gibberellins in the regulation of seed germination. Interestingly, it has a synergistic relationship with ethylene in stress conditions. The phytohormonal network is itself a complex mechanism where some hormones work synergistically and some are antagonistically to enhance the stress tolerance of suffering plants, as well as to enhance the growth of plants [30–32].

3. Transcriptional regulation

Activation of signal transduction networks also leads to the transcriptional regulation of plants. It follows the expression of stress-responsive genes that enhance the adaptive response of plants. For example, in response to drought stress, there is a modulation in the expression pattern of the WRKY, which further regulates the ABA-responsive genes that contribute toward the stomatal regulation and retention of water in stress conditions [33–35]. In salt stress, WRKY transcription factors regulate the ion transporter gene that scavenges ROS. In response to stress complex transcriptional response is transduced that regulates the ROS scavenging genes, chaperones, LEA proteins, and hormonal biosynthesis genes that play a significant role in plant stress tolerance [36, 37].

4. Conclusion and further prospects

Signal transduction plays an important role in the adaptation of plants' response to various stresses. The complex regulatory network of receptors, secondary messengers, and transcription factors enables the plant to adapt to stressful conditions. The stress perception occurs at the plasma membrane to metabolic reprogramming at the nucleus, and the whole cascade of events ensures the plant's adaptation to stress conditions. The knowledge of this complex transduction network enhances understanding to improve strategies for plant stress tolerance. Integrating recent advances in biotechnological intervention, modern breeding techniques can improve the signal transduction for enhanced stress tolerance. Future research on the identification of novel signaling components and the creation of transgenic plants with the addition of transcription factors that are involved in the transduction pathways should be explored.

Author details

Muhammad Waseem[1,2,3]

1 School of Breeding and Multiplication (Sanya Institute of Breeding and Multiplication) School of Tropical Agriculture and Forestry, Hainan University, Sanya, China

2 Fang Zhiyuan Academician Team Innovation Center, Hainan, China

3 Key Laboratory for Quality Regulation of Tropical Horticultural Crops of Hainan Province, Hainan, China

*Address all correspondence to: m.waseem.botanist@gmail.com; drwaseem@hainanu.edu.cn

References

[1] Sahu SK, Liu P, Chandrasekaran U, Waseem M. Salinity and drought stress in plants: Understanding physiological, biochemical and molecular responses, volume II. Frontiers in Plant Science. 2025;**16**:1625602

[2] Waseem M, Liu P, Aslam MM. Salinity and drought stress in plants: Understanding physiological, biochemical and molecular responses. Frontiers in Plant Science. 2023;**14**:1277859

[3] Jiang Z, van Zanten M, Sasidharan R. Mechanisms of plant acclimation to multiple abiotic stresses. Communications Biology. 2025;**8**(1):655

[4] Zhang H, Sonnewald U. Differences and commonalities of plant responses to single and combined stresses. The Plant Journal. 2017;**90**(5):839-855

[5] Mareri L, Parrotta L, Cai G. Environmental Stress and Plants. MDPI; 2022. p. 5416. Available from: https://www.mdpi.com/1422-0067/23/10/5416

[6] Boyer J. Environmental Stress and Crop Yields. Crop Reactions to Water and Temperature Stresses in Humid, Temperate Climates. CRC Press; 2019. pp. 3-7. Available from: https://www.routledge.com/Crop-Reactions-To-Water-And-Temperature-Stresses-In-Humid-Temperate-Climates/Kramer/p/book/9780367169084

[7] Kopecká R, Kameniarová M, Černý M, Brzobohatý B, Novák J. Abiotic stress in crop production. International Journal of Molecular Sciences. 2023;**24**(7):6603

[8] Kaur N, Gupta AK. Signal transduction pathways under abiotic stresses in plants. Current Science. 2005;**88**(11):1771-1780

[9] Palme K. Signals and Signal Transduction Pathways in Plants. Great Britain: Springer Science & Business Media; 2012

[10] Scheel D, Wasternack C. Plant Signal Transduction. Oxford University Press; 2002. Available from: https://www.amazon.com/Plant-Signal-Transduction-Dierk-Scheel/dp/0199638799

[11] Xiong L, Zhu JK. Abiotic stress signal transduction in plants: Molecular and genetic perspectives. Physiologia Plantarum. 2001;**112**(2):152-166

[12] Karuppanapandian T, Moon J-C, Kim C, Manoharan K, Kim W. Reactive oxygen species in plants: Their generation, signal transduction, and scavenging mechanisms. Australian Journal of Crop Science. 2011;**5**(6):709-725

[13] Kim DS, Kim J-B, Goh EJ, Kim W-J, Kim SH, Seo YW, et al. Antioxidant response of Arabidopsis plants to gamma irradiation: Genome-wide expression profiling of the ROS scavenging and signal transduction pathways. Journal of Plant Physiology. 2011;**168**(16):1960-1971

[14] Shaffique S, Farooq M, Kang S-M, Lee I-J. Recent advances in biochemical reprogramming network under drought stress in soybean. Journal of Soil Science and Plant Nutrition. 2024;**24**(2):1692-1703

[15] Taga T, Kishimoto T. Signaling mechanisms through cytokine receptors that share signal transducing receptor

components. Current Opinion in Immunology. 1995;7(1):17-23

[16] Torii KU. Leucine-rich repeat receptor kinases in plants: Structure, function, and signal transduction pathways. International Review of Cytology. 2004;**234**:1-46

[17] Geldner N, Robatzek S. Plant receptors go endosomal: A moving view on signal transduction. Plant Physiology. 2008;**147**(4):1565-1574

[18] Finkel T. Signal transduction by reactive oxygen species. Journal of Cell Biology. 2011;**194**(1):7-15

[19] Russell EG, Cotter TG. New insight into the role of reactive oxygen species (ROS) in cellular signal-transduction processes. International Review of Cell and Molecular Biology. 2015;**319**:221-254

[20] Tuteja N, Mahajan S. Calcium signaling network in plants: An overview. Plant Signaling & Behavior. 2007;**2**(2):79-85

[21] Batistič O, Kudla J. Analysis of calcium signaling pathways in plants. Biochimica et Biophysica Acta (BBA) – General Subjects. 2012;**1820**(8):1283-1293

[22] Sanders D, Jrm P, Brownlee C, Harper JF. Calcium at the crossroads of Signaling. The Plant Cell. 2002;**14**(Suppl. 1):S401-SS17

[23] Liu Y, He C. A review of redox signaling and the control of MAP kinase pathway in plants. Redox Biology. 2017;**11**:192-204

[24] Wu X, Liu Z, Liao W. The involvement of gaseous signaling molecules in plant MAPK cascades: Function and signal transduction. Planta. 2021;**254**(6):127

[25] Zeng H, Zhang Y, Wang S, Zhang X, Wang H, Du L. Calcium/calmodulin-mediated signal transduction system in plants. Chinese Bulletin of Botany. 2016;**51**(5):705-723

[26] Raja V, Majeed U, Kang H, Andrabi KI, John R. Abiotic stress: Interplay between ROS, hormones and MAPKs. Environmental and Experimental Botany. 2017;**137**:142-157

[27] Lv Z-Y, Sun W-J, Jiang R, Chen J-F, Ying X, Zhang L, et al. Phytohormones jasmonic acid, salicylic acid, gibberellins, and abscisic acid are key mediators of plant secondary metabolites. World Journal of Traditional Chinese Medicine. 2021;7(3):307-325

[28] Parwez R, Aftab T, Gill SS, Naeem M. Abscisic acid signaling and crosstalk with phytohormones in regulation of environmental stress responses. Environmental and Experimental Botany. 2022;**199**:104885

[29] Zheng L, Meng Y, Ma J, Zhao X, Cheng T, Ji J, et al. Transcriptomic analysis reveals importance of ROS and phytohormones in response to short-term salinity stress in Populus tomentosa. Frontiers in Plant Science. 2015;**6**:678

[30] Myers RJ Jr, Fichman Y, Zandalinas SI, Mittler R. Jasmonic acid and salicylic acid modulate systemic reactive oxygen species signaling during stress responses. Plant Physiology. 2023;**191**(2):862-873

[31] Chen H, Song L, Zhou H, Yao T, Zhang Z, Zhang H, et al. ABA signal transduction and ROS metabolic balance play a key role in the drought resistance of safflower. Plant Growth Regulation. 2025;**105**(1):273-294

[32] Arif Y, Sami F, Siddiqui H, Bajguz A, Hayat S. Salicylic acid in relation to other

phytohormones in plant: A study towards physiology and signal transduction under challenging environment. Environmental and Experimental Botany. 2020;**175**:104040

[33] Wani SH, Anand S, Singh B, Bohra A, Joshi R. WRKY transcription factors and plant defense responses: Latest discoveries and future prospects. Plant Cell Reports. 2021;**40**(7):1071-1085

[34] Javed T, Gao S-J. WRKY transcription factors in plant defense. Trends in Genetics. 2023;**39**(10):787-801

[35] Lee FC, Yeap WC, Appleton DR, Ho C-L, Kulaveerasingam H. Identification of drought responsive Elaeis guineensis WRKY transcription factors with sensitivity to other abiotic stresses and hormone treatments. BMC Genomics. 2022;**23**(1):164

[36] Rai GK, Mishra S, Chouhan R, Mushtaq M, Chowdhary AA, Rai PK, et al. Plant salinity stress, sensing, and its mitigation through WRKY. Frontiers in Plant Science. 2023;**14**:1238507

[37] Yu Y, Wu Y, He L. A wheat WRKY transcription factor TaWRKY17 enhances tolerance to salt stress in transgenic arabidopsis and wheat plant. Plant Molecular Biology. 2023;**113**(4):171-191

Chapter 2

Molecular and Physiological Responses of Plants to Salinity and Drought Stress: Implications for Crop Improvement and Climate Resilience

Rizwana Kanwal, Ansa Asghar, Usman Tufail, Asima Asghar, Mansoor Hameed, Amina Ameer, Zahida Parveen, Nida Hussain, Anjum Shaheen, Jazab Shafqat, Sana Basharat, Pingwu Liu and Muhammad Waseem

Abstract

Two of the most pervasive abiotic factors that pose a serious danger to global agricultural output and food security are salinity and drought. This chapter provides a comprehensive overview of the biochemical and physiological responses of plants to drought and salinity stress. These environmental challenges disrupt key processes such as cellular metabolism, photosynthesis, ion regulation, and water uptake, ultimately hindering plant growth and development. This chapter talks about how antioxidant systems, osmo-protectants, transcription factors, and signaling pathways can all improve stress tolerance. Interactions between drought and salinity reveal both overlapping and distinct pathways, while advances in omics approaches have identified critical genes and metabolic routes that underpin adaptation. Case examples from rice, wheat, and maize highlight the success of breeding programs and biotechnological strategies in enhancing stress resilience. The integration of molecular breeding, plant growth regulators, beneficial microorganisms, and nanotechnology further strengthens crop performance under stress. Overall, the chapter emphasizes that sustainable management of drought and salinity depends on a multi-faceted approach that blends conventional practices with modern innovations to build climate-resilient agriculture and ensure food security.

Keywords: salinity, drought, tolerance, mechanism, physiology, responses, nutrient balance, ROS homeostasis, antioxidant defenses, ABA-mediated signaling

1. Introduction

Salinity and drought are two of the most common abiotic conditions that significantly reduce the agricultural productivity globally [1]. The impact of drought and salinity stressors is further compounded by changes in the global climate that impact the frequency and severity of these stressors. To successfully address the issues posed by drought and salinity stressors in the current environment, innovative approaches to sustainable crop production are essential [2]. In light of this, considerable strides have been made in understanding the fundamental physiological, biochemical, and molecular underpinnings of salinity and drought stress; these discoveries have been extremely helpful in the breeding of crops that can withstand stress. Our comprehension of the body of knowledge about drought and salinity tolerance, however, continues to be a persistent problem. The main obstacles include (i) the development of resistance to salinity and drought, (ii) important genes in genetic and transcriptional and posttranscriptional regulatory mechanisms [3], (iii) the transfer of learned lessons from lab to field for sustainable production, and (iv) biochemical changes and metabolic adaptations to prevent cellular damage. Research projects that focus on the combined effects of drought and salinity stressors must be established in order to solve these issues. In order to establish the sustainable agriculture in the age of climate change, such research should concentrate on investigation at the genetic, molecular, and physiological levels.

2. Plant response mechanism to drought stress and salinity

Plants are often exposed to a wide range of abiotic stress conditions throughout their lifecycle [4]. Adapted organisms are those that have evolved to overcome such stressors and revert to normal basal metabolism by modulating multiple response systems in favor of their defense [5]. Crucially, agricultural productivity and expansion are strongly constrained by these environmental variables. For instance, a rise in atmospheric CO_2 may cause plants' photosynthetic rate to alter [6], resulting in a shift in growth rate that typically has a good effect on total biomass but a negative impact on nutritional quality. Stress-induced stimuli create specific alterations in the physiology and growth of plants. In this context, stress can alter a variety of systems, including lipid signaling, transposable elements, photosynthesis and gas exchange, cell death, alterations in the composition of the cell wall, nutrient translocation, gene transcription, metabolites, proteins, and antioxidant profiles illustrated in **Figure 1**.

When plants perceive these imbalances brought on by the abiotic pressures mentioned above, they shift their resources and energy toward defense and reproduction, which slows growth and lowers biomass production [7].

2.1 Water deficit of plants

Water availability has become a critical issue in the context of ongoing climate change and the rapidly growing global population, especially considering that agriculture accounts for over 70% of global freshwater usage [8]. Irrigated agriculture, which occupies only 20% of cultivated land, contributes to approximately 40% of global food production and is, on average, at least twice as productive per unit area as rainfed agriculture. Water is a fundamental component of plant biomass, typically constituting 50–90% of the fresh weight, with a significant portion retained within the cell walls (ranging from 60 to 90%) [9]. In this context, sustaining crop growth and productivity

DOI: http://dx.doi.org/10.5772/intechopen.1014325

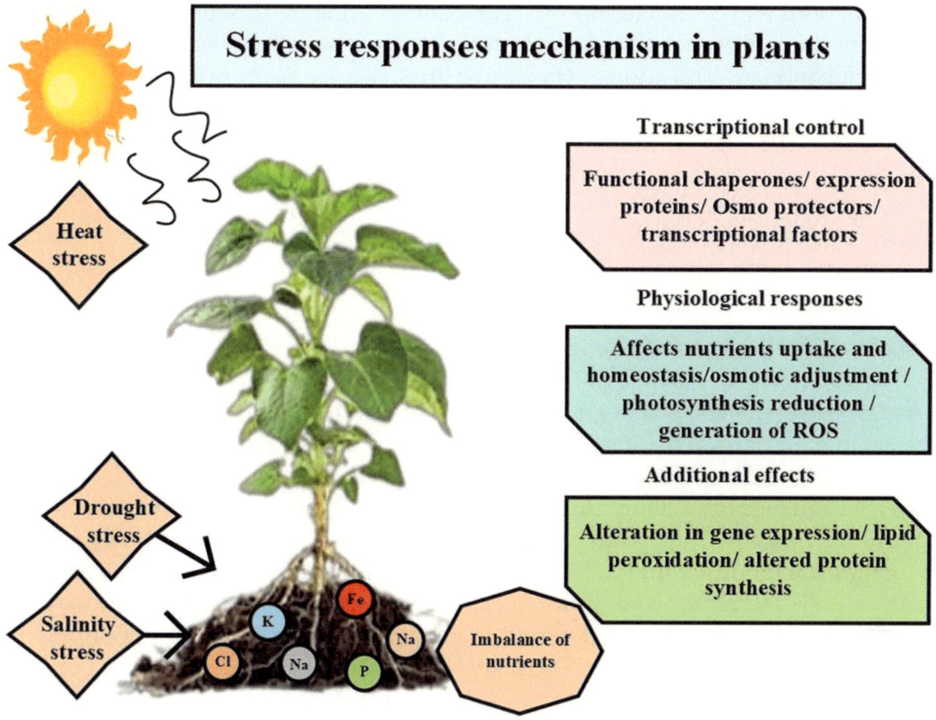

Figure 1.
Plant response cascade to different abiotic stresses (source: Author).

under adverse environmental conditions particularly water scarcity—remains one of the most pressing challenges in modern agriculture. A water deficit happens when a plant's regular functioning (such as growth, photosynthetic rate, and stomatal conductance) is impacted when its water demand cannot be entirely met (e.g. by adequate irrigation or precipitation). Stated differently, a water deficit arises when the rate of transpiration from the outermost layer of leaves surpasses the rate at which water is absorbed by the roots [10]. Drought, a significant abiotic stressor that adversely affects crop productivity and yield, is characterized as a period of unusually dry weather that causes a soil-water deficit and, in turn, a plant water deficit [11]. Climate change, rising food demand, and water resource depletion all worsen the effects of drought on agriculture. These are now the main constraints limiting crop output and, eventually, food security, which puts greater demands on farming to use freshwater more efficiently. Therefore, water use effectiveness (WUE), crop water consumption, and drought-tolerant plant development are essential for crop improvement [12]. WUE is the quantity of carbon that is absorbed as grain or biomass for every unit of water that the crop uses. One of the most important factors in plant preservation under drought stress is its ability to withstand moderate-to-severe soil water deficiency [13]. One of the main objectives of breeding and genetic engineering initiatives is to raise the WUE of crops. This region was linked to WUE and contained candidate genes, including glutathione S-transferase, abscisic acid-responsive element binding protein, late embryogenesis abundant proteins, and aquaporins [14]. Jia et al. [15] have shown that under long-term moderate drought circumstances, apple plants with overexpression of the MdATG8i-OE (autophagy protein) gene had higher WUE than wild-type plants.

There have been numerous attempts to create and/or enhance drought-tolerant plants throughout the past century. Numerous strategies have been employed to accomplish this goal, including the creation of transgenic lines, precise genome editing techniques, omics methodologies, and plant breeding [16]. Utilizing plant breeding techniques has the potential to significantly increase the production of drought-tolerant plants, and significant strides have been made in this area for a variety of crops, including barley, soybean, popcorn, rice, and wheat [17]. However, a limited genetic pool, poor heredity of dehydration tolerance/resistance, the intricacy of tolerance, and the magnitude of environmental interactions limit the capacity to create drought-tolerant plants.

2.2 Methods for addressing the water shortfall

As sessile organisms, plants are constantly exposed to a range of environmental stressors in their natural habitats [18]. Among these, water deficiency significantly impairs plant growth and development, ultimately leading to reduced yields. To cope with such challenges and maintain their water status, plants have evolved diverse adaptation and acclimation strategies at multiple levels, including morphological and anatomical modifications, as well as physiological, biochemical, and molecular responses [19]. The nature and duration of these responses are influenced by various factors, such as the severity and duration of water stress, plant species, developmental stage, and physiological maturity. The plant's collection of characteristics and essential functions gradually deteriorate over time, regardless of how brief the stress is [20]. While a severe water deficit may result in plant death, a moderate deficit generates notable morphological and physiological changes. This capacity of plants to endure or thrive in a tough environment brought on by stresses such as drought is known as drought resistance. This ability is a complex characteristic linked to several levels of adaptation that enables plants to adjust to particular environments for the sake of their development and growth [21]. The four categories of plant drought resistance strategies are escape, recuperation, tolerance, and avoidance. The preservation of elevated plant water potentials despite water scarcity is a hallmark of drought avoidance [22]. In addition to cellular metabolism (such as the formation of thicker networks of roots, stomatal closure, leaf rolling, and wax buildup on the leaf surface), it is in charge of morphological changes in plants. Drought tolerance depends on plants' capacity to maintain physiological functions in the face of extreme drought stress by reorganizing their metabolic and gene regulation systems to lessen or reverse the stress damage that results [23]. The term "drought escape" describes the synchronization of a plant's life cycle, growth phase, or planting timing in order to avoid drought throughout the growing season. The term "drought recovery" describes a plant's ability to regain its metabolism during a period of extreme drought, which results in turgor loss, leaf desiccation, and the entire growth cycle [24]. While brief but intense drought episodes promote drought tolerance, extended drought conditions are linked to drought avoidance or escape. As previously stated, plants have used a variety of tactics and at many levels from anatomical to molecular to adapt to water scarcity [25].

3. Physiological techniques for boosting output in water-scarce environments

Water scarcity reduces water-use efficiency, disrupts essential water-related processes, and impedes normal plant growth and development. In response, plants have evolved a range of morphological, physiological, and biochemical adaptations to mitigate the effects of drought stress [26]. These include alterations in transpiration rates, osmotic regulation, photosynthetic and respiratory activity, ion homeostasis, activation of antioxidant enzyme systems, and the expression of drought-responsive proteins [14].

The reactions of the leaves and roots at the level of morphology are essential for minimizing water loss and advancing WUE. In reaction to water-deficit stress, the root system is essential. Important characteristics linked to crop drought resilience include its length, weight, volume, and density [27]. Turgor pressure causes leaf rolling, a drought-adaptive characteristic that lowers water consumption during water stress. Stomata are pores that facilitate gas exchange between the environment and plants. They control the ideal transpiration while guaranteeing maximal transpiration of CO_2 for photosynthesis [28]. After root cells identify a water deficit in the soil, plants produce abscisic acid (ABA). The morphological and anatomical changes that result from ABA include senescence of the leaves, closure of stomatal pores, and improved plant water uptake from soil by lowering the water potential in contrast to the water-limiting environment through osmotic adjustments [29]. Zhu et al. [30] discovered that by primarily activating the ABA pathway, the transcription factor ipa1 (ideal plant architecture 1) greatly increased rice drought tolerance at the seedling stage In order to produce ipa1 plants with a more developed root system and a smaller leaf stomatal aperture, the authors employed a pair of IPA1 and ipa1-near isogenic lines. One rapid, short-term reaction to water restriction is ABA-mediated stomatal closure. Drought may enhance morpho-anatomical features, although physiological and biochemical constraints may affect plant performance more [31]. The centers of plant energy metabolism, photosynthesis and respiration, are adversely affected by dryness. As the relative water content of the soil decreases, photosynthetic and rates of transpiration also fall, making photosynthesis one of the primary processes impacted by water stress [32]. Stomatal closure limits internal CO_2 levels and carboxylation by reducing leaf CO_2 inflow. During mild-to-severe drought stress, the primary cause of the lowered photosynthetic rate is thought to be the reduced diffusion of CO_2 from the surrounding atmosphere to the carboxylation site. Additionally, water restriction modifies the electron transport chain (ETC), which in turn increases reactive oxygen species (ROS) (such as superoxide, hydrogen peroxide, and hydroxyl radicals) in cell organelles, disrupting thylakoid structures and adversely effecting PSI and PSII photosystems [33]. Since plants must use a great deal of energy to adapt to drought, it is an expensive phenomenon.

The ratio of photosynthetic CO_2 uptake to respiratory CO_2 emission determines plant growth. The equilibrium between respiration and photosynthesis determines how plant cells modify their carbon metabolism [6]. Photosynthetic rates are probably going to result in less substrate being supplied to the mitochondria, which will affect the rate of epidermal respiration. On the other hand, drought may make it more necessary for respiratory ATP to sustain cellular metabolism. ROS are created in the mitochondria of plants during drought and have the potential to harm various parts of the cell [34]. Alternative oxidase (AOX), a terminal oxidase that does not conserve energy, is a component of the plant mitochondrial ETC. Plants subjected to water scarcity can employ this mechanism and photorespiration to keep their cells functional by avoiding the buildup of ROS [31]. It is also possible to alter the TCA cycle pathway to stop the synthesis of excess reductants *via* the γ-aminobutyric acid (GABA) pathway. It builds up under stressful circumstances and could be an adaptive reaction to stress. It is also possible to alter the TCA cycle pathway to stop the synthesis of excess reductants *via* the γ-aminobutyric acid (GABA) pathway. It builds up under stressful circumstances and could be an adaptive reaction to stresses [35]. It is discovered that GABA does not cause changes in stomatal pore aperture; instead, it counteracts changes in pore size. They suggested that cytosolic GABA signals control stomatal opening, WUE, and drought resilience by negatively regulating the activity of ALMT9.

Inorganic acids, carbohydrates, polyamines, secondary metabolites, plant hormones, and amino acids are all essential components of stress tolerance systems [36]. The accumulation of specific compounds in plants under water stress plays a vital role in decreasing osmotic potential and enhancing cellular water retention. These compounds contribute to physiological responses by stabilizing membranes, regulating osmotic balance, scavenging reactive oxygen species (ROS), reducing ion leakage, limiting leaf area expansion, and promoting root growth [37]. A prominent example is the osmoprotectant amino acid proline, which functions as a critical signaling molecule during dehydration stress. Its accumulation helps maintain cellular redox balance and reduces ROS levels, thereby preserving membrane integrity through the mitigation of lipid peroxidation. It has been discovered that other amino acids contribute to drought tolerance. Akram et al. [38] discovered that methionine levels were higher in broccoli plants that could withstand drought. Under drought stress, cotton acquired proline, glutamate, glutarate, tryptophan, and valine. According to recent research, crops' tolerance to drought was enhanced by the exogenous administration of a variety of plant growth regulators (PGRs), including hormones, polyamines, organic acids, antioxidants, osmoprotectants, and others. According to these investigations, PGR treatment improves osmolyte and antioxidant accumulation, photosynthetic capability, relative water content (RWC), and gas exchange characteristics. In contemporary crops, this type of priming treatment has great promise for the creation of drought-resistant plants. As an illustration, Xie et al. [39] found that exogenous citric acid treatment in tobacco plants led to decreased conductance of stomatal pores, transpiration, and water loss during drought conditions and increased chlorophyll content, net the cycle of photosynthesis relative water content, and abscisic acid content.

4. Challenges and developments in drought-tolerant cultivar development

Drought tolerance is not a simple qualitative trait but rather a complex, quantitative polygenic trait governed by multiple genes, each exerting a relatively small effect [40]. The development of drought-tolerant cultivars is further complicated by low heritability, strong genotype-by-environment (G × E) interactions, and the challenges associated with large-scale screening of drought-responsive genes and traits. This intricacy explains the slow development in yield enhancement in drought-prone regions and the challenge of comprehending the molecular pathways triggered by drought. Some specific crops now have genetically enhanced drought-tolerant lines or cultivars thanks to traditional plant breeding [41]. Nevertheless, this method is very labor-, time-, and money-intensive. Drought tolerance processes have been better understood in the recent decades thanks to advances in agricultural physiology, marker-assisted breeding, and genomics, which have also produced new information and tools for plant development. Some of the latest technologies are based on omics methodologies (genomics, transcriptomics, proteomics, and metabolomics), gene editing techniques, genome-wide selection (GWAS), and marker-assisted selection (MAS) [42].

Although genome-wide association studies (GWAS) and quantitative trait loci (QTL) mapping are valuable tools for identifying genomic regions and alleles associated with drought-related traits, our understanding of the specific candidate genes and their functional roles remains limited. Integrating high-throughput phenotyping with multi-omics approaches, such as genomics, transcriptomics, proteomics, and metabolomics alongside modern breeding strategies, holds great promise for enhancing our knowledge of the genetic and physiological mechanisms underlying crop responses to drought stress (**Figure 2**) [43].

4.1 Salinity

Over the past five decades, land salinization has emerged as a major constraint in agriculture [44]. Currently, nearly one-third of the world's arable land is affected by salinity, leading to substantial declines in crop productivity and overall agricultural output. Poor irrigation techniques, improper use of agrochemicals, and industrial pollution all contribute to the global aggravation of soil salinization [45]. Meanwhile, general salt concentrations can alter physiological and biochemical processes, limiting the development of the plant's aerial portion as well as its root system. Plants are stressed by salt in two ways: osmotically and ionically. Restricted transpiration by plant roots results from high levels of salt in the soil because they lower the soil's water potential [46]. Conversely, an overabundance of ions such as Na^+ and Cl^- in cells causes nutritional problems or plant toxicity. In this regard, it is essential to enhance the management strategies to lessen the harm brought about by this phenomenon and to step up research and plant development initiatives targeted at boosting salt stress

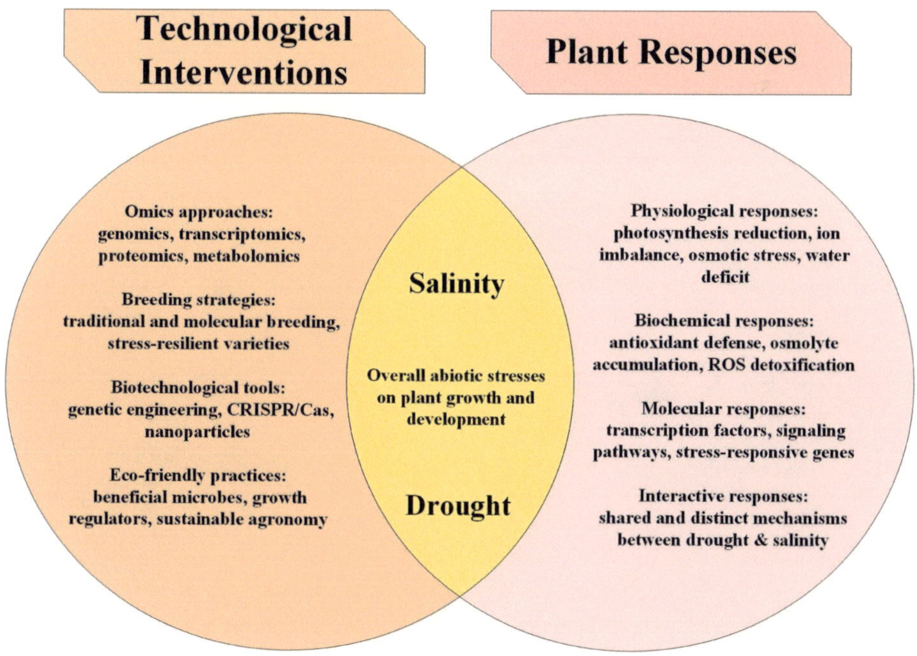

Figure 2.
Plant response, and technological interventions of plants during drought and salt stress (source: Author).

resistance [47]. Given the forecasts of population expansion and climate change, agricultural land salinization presents a serious concern for the future.

4.2 Response mechanisms to saline stress

An osmotic compensation mechanism is triggered in plants exposed to salinity in order to preserve the turbidity of the cells; this causes stressed plants to grow slowly [48]. Meanwhile, the alterations that salt stress causes in plants might differ depending on the genotype, developmental stage, stress level, and duration, as well as the management. To cope with the associated stress, plants can initiate several physiological and biochemical processes (**Figure 3**), such as morphological, anatomical, water relations, photosynthesis, hormones, ion distribution, and biochemical adaption modifications [49].

The two primary phases of a plant's response to salinity are (i) the osmotic phase, which is characterized as rapid because the salt lowers the osmotic capacity of the soil–plant interactions as it reaches the roots [50], resulting in a decrease in shoot growth because of the decreased water potential; and (ii) the slow phase, which involves the absorption and indicating of toxic ions like Na+. Since cell homeostasis is achieved through the interaction of these ions, cations (Na+, Ca2+, Mg2+, and K+) and anions (Cl−, SO4−, HCO3−, CO32−, and NO3−) are the primary ions implicated in salt stress signaling [51]. Extreme salinization causes plants to either restrict the entry of Na + or store it in their mature tissues, creating a "stock" environment that will eventually be removed. Halophyte plants are defined as such because they have evolved traits and systems that can adapt to the saltwater environment [52]. Numerous thorough reviews covering different facets of halophyte physiology have been published. In order to cope with the stress of saline soils, halophyte plants have evolved three primary strategies: absorbing high concentration ions and storing

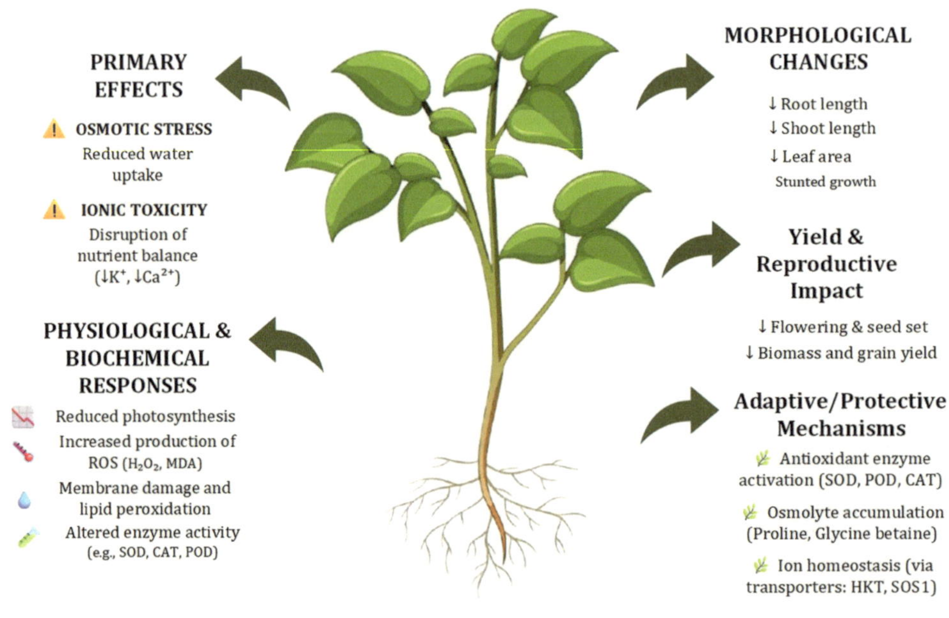

Figure 3.
Overview of key physiological, biochemical, and molecular effects induced by salt stress in plants (source: Author).

them in vacuoles; specific cells removing absorbed salts; and stem cells limiting salt uptake [53]. Glycophytes, or plants that are sensitive to salinity, can collect very little Na + and Cl−. Because of poor salt compartmentation, glycophytes are more vulnerable to the buildup of these ions in their tissues, which can impair growth and productivity. When plants are exposed to saline circumstances, the roots are typically the first organ to detect the salt signal. The inhibition of proliferation of cells, initiation, extension, or developmental redirection away from salt mediates the root architectural changes in response to salinity [54]. Additionally, plants can build tolerance by altering root growth processes that necessitate metabolic changes in order respond to stressful situations. Farooq et al. [54] demonstrated in this regard that salt-adapted cultured cells from Arabidopsis roots produced greater amounts of carbohydrates, amino acids, and intermediate metabolites in the shikimate pathway, including coniferin, in callus suspension. The necessity of modifying physical characteristics during adaptation to high-saline circumstances is further supported by the fact that adapted cells developed thicker cell walls with higher lignin levels [55]. An rise in the root-shoot ratio, which is linked to water stress rather than a particular effect of salt, is another common reaction to salt stress. Under the conditions of salt stress, a larger proportion of roots may enhance the retention of harmful ions in this organ, thereby regulating their transfer to the aerial portions [56]. Furthermore, because plants are more likely to explore the soil, a high spatial distribution of roots under salt stress may benefit water absorption.

In plants, the photosynthetic process affects nutrient fluxes and productivity. The primary cause of decreased plant productivity is salinity's impact on the physiological functions of the leaf, specifically photosynthesis [57]. Salinity exposure causes stomatal closure, which lowers the photosynthetic ratio because stomatal conductance decreases, limiting CO_2 access for the Calvin-Benson cycle. Furthermore, salinity may affect photosynthesis by destroying the oxygen-evolving complex (OEC), inhibiting electron transport, inactivating the PSII reaction centers, and reducing the donor side of PSII's electron transfer capability [58].

In plants, salinity causes oxidative stress at the subcellular level. When oxygen is decreased by reducing molecules, ROS are created. Stress from the environment causes their concentration to rise, and the stomata close to stop water loss [59]. The reducing power of NADPH, or reduced ferredoxin, is transferred to oxygen by the reduction of CO_2 in the leaves. As a result, oxygen is reduced in its radical forms, including superoxide (O_2 •−), hydrogen peroxide (H_2O_2), and hydroxyl radical ($HO^{\cdot}$). Proteins, nucleic acids, and membrane lipids are oxidatively damaged as a result of the elevated ROS levels in plant tissues. Plants create an effective system of enzymatic and non-enzymatic antioxidants to scavenge the elevated ROS levels [60]. Reduced ascorbate (ASC), glutathione (GSH), phenolics, flavonoids, and tocopherols are examples of non-enzymatic antioxidants.

Antioxidants were found to be up-regulated in a variety of cultivated plants, including tomato, soybean, maize, wheat, and rice [61]. Salt-tolerant cultivars showed that a more effective antioxidative system is frequently linked to a greater resistance to salinity, which in turn reduces oxidative damage. Furthermore, numerous studies have shown the importance of both antioxidants with and without enzymes in medicinal plants such as *Curcuma longa* L., *Melissa officinalis* L., *Hyssopus officinalis* L., and *Lonicera japonica* L. This demonstrates how antioxidants can reduce oxidative damage brought on by salt stress [62].

One of the most significant changes brought on by the overabundance of these ions is the plant's decreased ability to absorb water, which can lead to a decrease in the absorption of nutrients due to ion toxicity [63]. In this manner, a rise in salinity caused by salt stress results in a potassium (K) deficit. Stress causes Na + to build up in salt, which is followed by a precipitous drop in K+ level. The rivalry between Na + and K+ uptake in the roots is the cause of this negative relationship. For instance, a decrease in plant hydration status, restricted uptake of macronutrients (N, P, and K) [64], and a decreased K+/Na + ratio all contributed to decreased growth and production in wheat cultivars under salt stress (150 mM). Furthermore, salinity can affect the various stages of nitrogen (N) metabolism, including protein synthesis, absorption, and assimilation, by interfering with plants' ability to acquire and use N [65]. According to Alsaeedi et al. [66], salt stress dramatically slowed down cucumber leaves' ability to absorb nitrogen. In this way, under stressful salinity circumstances, the nutritional imbalance helps to lower the growth rate.

4.3 Molecular approach's to salt stress

Decoding the genome sequences of major food crops provides valuable insights for fundamental, functional, and comparative genomics, facilitating the identification of novel traits and the genetic improvement of cultivated varieties [67]. For example, the identification of quantitative trait loci (QTLs) associated with salt tolerance represents a critical step toward the development of salt-tolerant cultivars, which can enhance crop productivity in saline environments. Thus, many QTLs linked to adaptations to salt have been found in rice, maize, and soybeans, three cultivated crops [68]. The QTLs were mostly found for a number of agronomic features, such as physiologically and agronomic traits related to root architecture, yield and factors that contribute to yield, and salinity tolerance.

Understanding the alterations in the expression of genes, proteins, and metabolites that occur in response to salt exposure can be achieved by combining several omics techniques [69]. In order to comprehend the process of salt stress tolerance and its potential application to enhance plant tolerance to salinity, research has been conducted in the fields of genomes, transcriptomics as well as proteomics, metabolomics, and epigenomics [69]. With the advancement of numerous cutting-edge technologies, including as CRISPR/Cas9, target-site genome editing has become possible, allowing for the creation of salt-tolerant cultivars. For rice and tomatoes, some contemporary examples were created [70]. There are further reports of molecular methods utilized to comprehend and create salt-tolerant plants.

Important transcription factors (TFs) such as bZIP, DREB, MYC, MYB, NAC, and WRKY play crucial roles in salinity stress responses and hold significant potential for improving crop tolerance to salt stress [71]. For instance, Wang et al. reported that overexpression of the ThNAC12 transcription factor in *Tamarix hispida* enhanced salt tolerance by regulating reactive oxygen species (ROS) scavenging through the direct activation of ThPIP2;5 expression. Similarly, overexpression of SiMYB19 in transgenic rice led to increased abscisic acid (ABA) synthesis and upregulation of OsPK1 and OsABF2, genes associated with the ABA receptor signaling pathway [71]. It also encourages the accumulation of ABA. The CRISPR-Cas9 system in conjunction with overexpression technology demonstrates that GmNAC06 TF may induce the build-up of the amino acid proline and glycine betaine in order to mitigate or prevent the adverse effects of ROS under salt stress; in a similar manner, it may regulate the Na+/K+ ratios in root hairs in order to preserve ionic homeostasis in *Glycine max*. This is how GmNAC06 responds to salt exposure and may help create transgenic crops that can withstand salt. The creation of these new techniques will open up new possibilities for utilizing and investigating these instruments to enhance the agricultural characteristics [72].

5. Omics and advanced tools in stress research

Plant stress biology has been revolutionized by the quick development of high-throughput technologies, which have made it possible for researchers to decipher the intricate processes governing how plants react to abiotic stresses like drought and salinity [73]. Decoding multilayered stress responses at several molecular levels is made possible by omics platforms, which include transcriptomics, proteomics, metabolomics, ionomics, and genomes shown in **Figure 4**. These systems create new possibilities for creating crops that can withstand climate change when combined with genome editing techniques like CRISPR/Cas9 [74]. These methods for studying plant stress tolerance not only advance basic understanding but also make precision breeding and genetic engineering techniques for stress-resilient cultivars easier (**Table 1**).

5.1 Transcriptomics and proteomics

Transcriptomics encompasses the comprehensive study of RNA transcripts generated from the genome under the defined environmental or physiological conditions [82]. Under drought and salinity stress, transcriptomic analyses allow for the identification of thousands of differentially expressed genes (DEGs), which encode proteins associated with key stress-response pathways. These include genes responsible for osmotic regulation, antioxidative defense, signal transduction (e.g., MAPK cascades), transcription factors (such as DREB, MYB, WRKY), and stress-induced hormones like abscisic acid (ABA) [83]. Modern tools such as RNA-Seq and microarrays have made it possible to monitor temporal and tissue-specific gene expression with great precision, offering detailed insights into how plants adapt their transcriptome to mitigate damage.

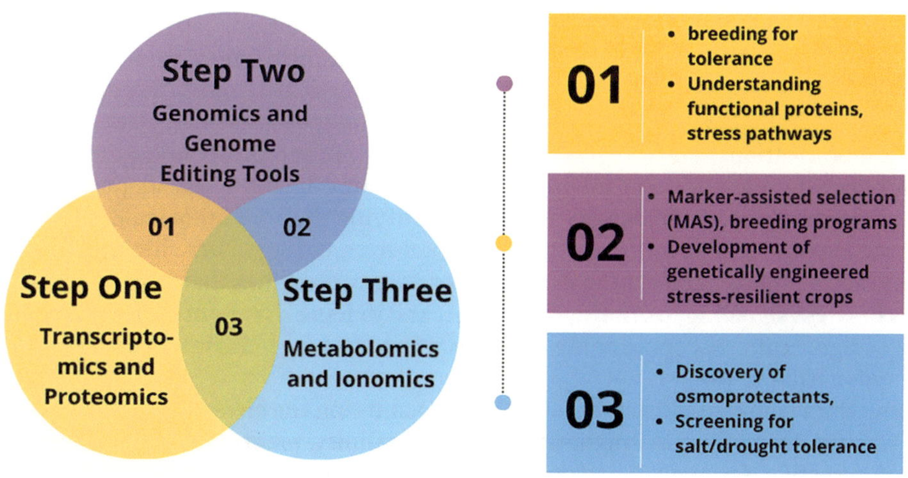

Figure 4.
Omics and advanced tools in stress research (source: Author).

Omics tool/ technology	Focus area	Key insights under salinity/drought stress	Applications	Reference
Transcriptomics	Gene expression profiling	Identification of stress-responsive genes (e.g., DREB, WRKY, MYB, NAC)	Marker discovery, breeding for tolerance	[75]
Proteomics	Protein abundance and function	Detection of stress-induced proteins (e.g., HSPs, LEA, antioxidant enzymes)	Understanding functional proteins, stress pathways	[76]
Metabolomics	Small molecule profiling	Accumulation of metabolites such as proline, glycine betaine, trehalose	Discovery of osmoprotectants, metabolic pathway analysis	[77]
Ionomics	Elemental composition and distribution	Ion homeostasis under stress; reduced Na^+ uptake, increased K^+ and Ca^{2+} balance	Screening for salt/ drought tolerance based on ion profiles	[78]
Genomics	Whole genome analysis	Mapping of QTLs and identification of functional markers	Marker-assisted selection (MAS), breeding programs	[79]
Genome editing (CRISPR/Cas9)	Targeted gene modification	Precise editing of stress-regulating genes; knockout of negative regulators	Development of genetically engineered stress-resilient crops	[80]
Transgenic approaches	Gene transfer across species	Overexpression of tolerance genes like DREB1A, NHX1	Improved stress tolerance in model and crop plants	[81]

Table 1.
Omics and advanced tools in salinity and drought stress research (source: Author).

Proteomics, on the other hand, deals with the large-scale study of the complete protein set within a plant cell or tissue under specific conditions [84]. While transcriptomics reveals which genes are being transcribed, proteomics provides information about which proteins are actually synthesized, modified, and active during stress conditions. Drought and salinity affect protein abundance, phosphorylation, and redox modifications. The integration of transcriptomic and proteomic data provides a comprehensive view of gene-to-function relationships during stress adaptation [85]. Metabolomics focuses on identifying and quantifying the small-molecule metabolites present in plants under different environmental stresses [86]. These include sugars, amino acids, organic acids, polyols, and secondary metabolites that play important roles in stress signaling, osmotic adjustment, and defense. Drought and salinity often trigger significant reprogramming of primary metabolism to maintain energy supply and redox balance [87].

Ionomics complements metabolomics by examining the elemental composition and ion distribution within plant tissues. Under salinity, plants face ionic imbalance due to excessive Na^+ and Cl^- accumulation, which disrupts nutrient uptake and metabolic functions [88]. Maintaining K^+/Na^+ homeostasis is critical for salt tolerance, and this is regulated by ion transporters such as HKT1, SOS1, and NHX [89]. Drought stress also influences the uptake of essential nutrients like nitrogen, phosphorus, and micronutrients due to reduced transpiration and root activity. Ionomic analyses using techniques such as inductively coupled plasma mass spectrometry (ICP-MS) enable the quantification of macro- and micro-elements in stressed plants, aiding in the identification of tolerant cultivars that maintain better nutrient and ion balance under adverse conditions [90].

5.2 Genomics and genome editing tools

Genomic approaches have significantly advanced the identification of genes, regulatory sequences, and chromosomal regions associated with abiotic stress tolerance [75]. Through whole-genome sequencing, genome-wide association studies, and quantitative trait loci mapping, researchers have pinpointed numerous candidate genes and loci related to drought and salinity tolerance in major crops like rice, wheat, maize, and soybean [91]. For instance, QTLs associated with root depth and ion exclusion are critical for breeding programs. In addition, high-throughput genotyping and molecular markers (SSR, SNPs) are now widely used for marker-assisted selection (MAS), allowing the rapid development of stress-resilient cultivars with high precision [92].

One of the most revolutionary breakthroughs in plant science has been the application of genome editing tools such as CRISPR/Cas9 [93]. This system enables targeted mutagenesis or gene insertion, offering unparalleled specificity and efficiency in modifying key genes involved in stress responses. For example, CRISPR has been successfully used to knock out negative regulators of drought responses or enhance the function of salt tolerance genes. Unlike traditional transgenic approaches, CRISPR allows precise edits without introducing foreign DNA, thus avoiding public and regulatory concerns associated with GMOs [94]. Additionally, genome editing can be combined with other omics data to validate gene function and fine-tune the expression of stress-responsive networks.

Together, genomics and genome editing tools provide a platform for accelerating genetic improvements in crops by directly targeting the most promising candidate genes [95]. The integration of omics with CRISPR-based precision breeding offers a holistic strategy for engineering plants that can thrive under the growing challenges posed by climate change and environmental stressors.

6. Strategies for enhancing stress tolerance

Enhancing plant tolerance to salinity and drought stress requires a multifaceted approach that integrates traditional and modern technologies [96]. One of the most longstanding strategies is conventional breeding, which involves the selection and cross-breeding of genotypes exhibiting desirable traits such as deep root systems, osmotic adjustment, or salt exclusion. While time-consuming, this method has been instrumental in developing stress-resilient cultivars in crops like rice, wheat, and maize [96]. Molecular breeding accelerates the development of tolerant cultivars by enabling early and accurate selection of stress-related traits.

The deployment of stress-tolerant cultivars developed through breeding programs is a practical and environmentally friendly strategy to ensure crop productivity in stress-prone regions [97]. These cultivars often possess multiple adaptive traits such as efficient water usage, salt exclusion mechanisms, and stable yield performance under stress conditions. For example, salt-tolerant rice lines like "Pokkali" or drought-tolerant maize hybrids have shown substantial improvements in marginal environments. Still, to overcome the limitations of naturally occurring genetic variation, scientists have turned to genetic engineering and biotechnological interventions [98]. Transgenic approaches allow for the introduction of specific genes that enhance stress responses, such as DREB, NHX, or HVA1. These genes can regulate transcription, ion transport, or osmoprotectant synthesis, offering precise control over the plant's response to stress conditions.

Emerging strategies also focus on the application of plant growth regulators (PGRs), beneficial microbes, and nanoparticles to mitigate the stress effects [99]. PGRs such as abscisic acid, salicylic acid, and brassinosteroids modulate physiological processes like stomatal closure, antioxidative defense, and root development. In parallel, plant growth-promoting rhizobacteria (PGPR) and arbuscular mycorrhizal fungi (AMF) improve nutrient uptake, enhance root architecture [100], and induce systemic tolerance by modulating hormonal signaling and activating defense genes. Recent advances in nanotechnology have introduced nanoparticles (e.g., ZnO, Si, TiO_2) as the promising tools to improve plant resilience [101]. These nanoparticles can enhance seed germination, stimulate antioxidant defense, and reduce cellular damage under stress through controlled release of nutrients and interaction with signaling pathways. Collectively, these strategies represent a synergistic framework for improving plant stress tolerance. By combining genetic, molecular, and physiological interventions with external applications of supportive agents, it is possible to create robust crop varieties that can withstand the increasing frequency and intensity of abiotic stress caused by climate change [102].

6.1 Case studies and examples of stress-tolerant genotypes

The development and deployment of stress-tolerant genotypes have played a crucial role in sustaining crop production under increasingly variable environmental conditions [97]. Numerous crop breeding programs worldwide have successfully identified and commercialized cultivars that exhibit improved tolerance to salinity and drought, offering valuable case studies for understanding the mechanisms of plant resilience [103].

In rice (*Oryza sativa*), one of the most salt-sensitive cereal crops, the traditional landrace "Pokkali," cultivated in the saline coastal regions of Kerala, India, has shown natural tolerance to high salinity levels [104]. This variety maintains low sodium accumulation in shoots through efficient Na^+ exclusion and compartmentalization mechanisms. Leveraging this trait, researchers developed improved lines like "FL478," which carries the Saltol QTL responsible for Na^+/K^+ homeostasis and exhibits high salinity tolerance at the seedling stage [104]. In the case of drought tolerance, the "Sahbhagi Dhan" rice variety, developed by the International Rice Research Institute (IRRI), combines deep rooting traits, early maturity, and osmotic adjustment capacity to perform well under limited water availability [105].

In wheat (*Triticum aestivum*), the genotype "Kharchia 65" is widely recognized for its salinity tolerance [106]. It accumulates lower levels of Na^+ in leaves and maintains high K^+/Na^+ ratios, enabling continued photosynthetic activity under saline stress. This genotype has been used in breeding programs to transfer salt tolerance traits into high-yielding backgrounds. For drought-prone environments, varieties such as "C306" and "KRL 210" demonstrate high yield stability and superior physiological traits like high relative water content (RWC), stay-green phenotype, and efficient transpiration regulation [107].

In maize (*Zea mays*), a moderately drought-sensitive crop, significant progress has been made in developing stress-tolerant hybrids through both conventional and molecular breeding [107]. The "DTMA" (Drought-Tolerant Maize for Africa) initiative, led by CIMMYT and partners, has successfully released more than 200 drought-tolerant varieties. These hybrids possess traits such as early flowering, deeper root systems, and improved water use efficiency [108]. One notable example is the hybrid "Zea DT CML442," which maintains yield stability under both moderate and severe water stress.

These case studies underscore the value of integrating physiological screening, molecular tools, and traditional knowledge to identify and promote genotypes capable of thriving under salinity and drought conditions. Their success not only contributes to food security in stress-affected regions but also serves as a blueprint for future breeding efforts in climate-resilient agriculture.

6.2 Future perspectives

Looking forward, enhancing plant resilience to salinity and drought stress requires continued innovation and global collaboration. Future research should focus on the integration of multi-omics platforms combining transcriptomics, proteomics, metabolomics, and phenomics to generate a holistic view of plant stress responses. Precision breeding and genome editing will play a central role in customizing crops for specific environments, especially when coupled with artificial intelligence and machine learning to predict gene function and phenotypic outcomes. Moreover, climate-smart agriculture should incorporate real-time monitoring, remote sensing technologies, and decision-support systems for optimal resource use. Sustainable practices like the use of bioinoculants, stress-resilient microbial consortia, and environmentally safe nanomaterials need to be scaled up for field-level applications. Strengthening public-private partnerships, farmer engagement, and policy support will also be essential for translating lab-scale advances into widespread agricultural impact. Ultimately, the goal is to ensure global food security by developing crops that are not only high-yielding but also capable of withstanding the increasing frequency of environmental extremes.

6.3 Limitations and real-world challenges

Finding genes, pathways, or physiological characteristics that are sensitive to stress and successfully implementing them into crop enhancement initiatives provide another difficulty. Although genetic engineering and molecular breeding techniques have potential, actual implementation is frequently hampered by problems such genotype-environment interactions, the stability of added characteristics, and the socioeconomic and regulatory obstacles to the adoption of genetically modified crops. The deployment of cutting-edge technology is further limited by resource constraints in the underdeveloped nations, where drought and salinity challenges are most acute.

7. Conclusion

Drought and salinity are among the most significant abiotic stresses that adversely affect plant growth and limit agricultural productivity on a global scale. These stresses disrupt various physiological, biochemical, and molecular processes, leading to reduced photosynthesis, nutrient imbalance, osmotic stress, oxidative damage, and impaired development. Understanding the complex molecular and physiological responses of plants has become essential to mitigate these effects. Plants exhibit a wide range of adaptive responses including osmolyte accumulation, antioxidant defense, ion transport regulation, and activation of specific signaling pathways. The integration of omics technologies, advanced breeding techniques, and agronomic practices has significantly improved our ability to dissect these responses and enhance stress tolerance. The application of plant growth regulators, beneficial microbes, and nanotechnology, along with genome editing tools like CRISPR/Cas9, has added precision and efficiency to current strategies. Altogether, a multi-disciplinary and integrative approach is needed to develop climate-resilient crops capable of thriving under harsh environments.

Abbreviations

APX	ascorbate peroxidase
MDHAR	monodehydroascorbate dehydrogenase
DHAR	dehydroascorbate reductase
GR	glutathione reductase
RWC	relative water content
PGRs	plant growth regulator
CK	cytokinin
GABA	γ-aminobutyric acid
QTL	quantitative trait loci
GSH	glutathione

Author details

Rizwana Kanwal[1], Ansa Asghar[1], Usman Tufail[1], Asima Asghar[2], Mansoor Hameed[1], Amina Ameer[1], Zahida Parveen[1], Nida Hussain[1], Anjum Shaheen[1], Jazab Shafqat[1], Sana Basharat[3,4], Pingwu Liu[3,4]* and Muhammad Waseem[3,4]*

1 Department of Botany, University of Agriculture Faisalabad, Pakistan

2 Department of Zoology, Lahore College of Women University, Lahore, Pakistan

3 School of Breeding and Multiplication, Sanya Institute of Breeding and Multiplication, Sanya, China

4 School of Tropical Agriculture and Forestry, Hainan University, Sanya, China

*Address all correspondence to: hnulpw@hainanu.edu.cn
and m.waseem.botanist@gmail.com

References

[1] Rahim HU, Ali W, Uddin M, Ahmad S, Khan K, Bibi H, et al. Abiotic stresses in soils, their effects on plants, and mitigation strategies: A literature review. Chemistry and Ecology. 2025;**41**(4):552-585

[2] Farhaoui A, Kouighat M, Khadiri M, Boutagayout A, Assouguem A, El Jarroudi M, et al. Innovative approaches to alleviate climate stress in crop production. In: Innovations in Climate Resilient Agriculture. Cham: Springer; 2025. pp. 271-307

[3] Hussain Q, Asim M, Zhang R, Khan R, Farooq S, Wu J. Transcription factors interact with ABA through gene expression and signaling pathways to mitigate drought and salinity stress. Biomolecules. 2021;**11**(8):1159

[4] Zhang H, Zhu J, Gong Z, Zhu J-K. Abiotic stress responses in plants. Nature Reviews Genetics. 2022;**23**(2):104-119

[5] Gorr TA. Hypometabolism as the ultimate defence in stress response: How the comparative approach helps understanding of medically relevant questions. Acta Physiologica. 2017;**219**(2):409-440

[6] Dusenge ME, Duarte AG, Way DA. Plant carbon metabolism and climate change: Elevated CO_2 and temperature impacts on photosynthesis, photorespiration and respiration. New Phytologist. 2019;**221**(1):32-49

[7] Nawaz M, Sun J, Shabbir S, Khattak WA, Ren G, Nie X, et al. A review of plants strategies to resist biotic and abiotic environmental stressors. Science of the Total Environment. 2023;**900**:165832

[8] Mishra RK. Fresh water availability and its global challenge. British Journal of Multidisciplinary and Advanced Studies. 2023;**4**(3):1-78

[9] Huang C, Zhou X, Zhan Y, Meng X, Fang G, Ling Z, et al. Evoking research attention back on liquid hot water pretreatment: A comprehensive review. Green Chemistry. 2025;**27**:4408-4422

[10] Suslov M. The dynamics of cell-to-cell water transport and the involvement of Aquaporins in response to Apoplast blockage in the roots of intact maize plants. Cells. 2025;**14**(12):902

[11] Jain S, Singh A, Prakash J. Drought and heat stress in fruit crops: Implications, mitigation, and policy perspectives. In: Drought and Heat Stress in Agriculture: Implications, Mitigation and Policy Approaches. Singapore: Springer; 2025. pp. 69-91

[12] Thakur H, Nishant, Mansi, Barkha. Enhancing water use efficiency in pulse-based cropping systems: A review. International Journal of Research in Agronomy. 2025;**8**(5):608-612. DOI: 10.33545/2618060X.2025.v8.i5h.2957

[13] Khan N, Ali S, Tariq H, Latif S, Yasmin H, Mehmood A, et al. Water conservation and plant survival strategies of rhizobacteria under drought stress. Agronomy. 2020;**10**(11):1683

[14] Haghpanah M, Hashemipetroudi S, Arzani A, Araniti F. Drought tolerance in plants: Physiological and molecular responses. Plants. 2024;**13**(21):2962

[15] Jia X, Mao K, Wang P, Wang Y, Jia X, Huo L, et al. Overexpression of MdATG8i improves water use efficiency in transgenic apple by

modulating photosynthesis, osmotic balance, and autophagic activity under moderate water deficit. Horticulture Research. 2021;**8**:81. DOI: 10.1038/s41438-021-00521-2

[16] Kumar A, Anju T, Kumar S, Chhapekar SS, Sreedharan S, Singh S, et al. Integrating omics and gene editing tools for rapid improvement of traditional food plants for diversified and sustainable food security. International Journal of Molecular Sciences. 2021;**22**(15):8093

[17] Begna T. Breeding strategies for improvement of drought tolerant in crops. World Journal of Agricultural Sciences. 2022;**18**(3):177-184

[18] Abdelsalam SS, Mugwanya M, Grossi CE, Bwambale J. Plant small peptides: Drivers of plant-stress tolerance, phytoremediation and diversity of root-associated microbes. Discover Plants. 2025;**2**(1):1-25

[19] Zahedi SM, Karimi M, Venditti A, Zahra N, Siddique KH, Farooq M. Plant adaptation to drought stress: The role of anatomical and morphological characteristics in maintaining the water status. Journal of Soil Science and Plant Nutrition. 2025;**25**(1):409-427

[20] Durgadevi P, Girigoswami K, Girigoswami A. Biodegradable nanomaterials in boosting seed vigor and germination: Seed coating towards sustainability. Discover Applied Sciences. 2025;**7**(7):695

[21] Kortman VG, Mazzolai B, Sakes A, Jovanova J. Perspectives on intelligence in soft robotics. Advanced Intelligent Systems. 2025;**7**(1):2400294

[22] Šola I, Gmižić D, Miškec K, Ludwig-Müller J. Impact of water stress on metabolic intermediates and regulators in broccoli sprouts, and cellular defense potential of their extracts. International Journal of Molecular Sciences. 2025;**26**(2):632

[23] Sarma H, Ramchiary C, Sen B, Daimary M, Prasad R. Integration of nitric oxide signaling in plant stress responses: Unveiling its role in enhancing crop tolerance to abiotic stress. Discover Plants. 2025;**2**(1):1-19

[24] Liu J, Zhao J, Ling X, Xiong D. 13C labeling to determine intra-leaf photosynthetic heterogeneity dynamics during drought and rewatering. Journal of Experimental Botany. 9 Oct 2025;**76**(16):4640-4653. DOI: 10.1093/jxb/eraf215

[25] Chachar S, Ahmed N, Hu X. Drought-induced aesthetic decline and ecological impacts on ornamentals: Mechanisms of damage and innovative strategies for mitigation. Plant Biology. 2025. Available from: https://onlinelibrary.wiley.com/doi/10.1111/plb.70074?af=R#

[26] Wach D, Skowron P. An overview of plant responses to the drought stress at morphological, physiological and biochemical levels. Polish Journal of Agronomy. 2022;**50**:25-34

[27] Ranjan A, Sinha R, Singla-Pareek SL, Pareek A, Singh AK. Shaping the root system architecture in plants for adaptation to drought stress. Physiologia Plantarum. 2022;**174**(2):e13651

[28] Devi MJ, Reddy VR. Transpiration response of cotton to vapor pressure deficit and its relationship with stomatal traits. Frontiers in Plant Science. 2018;**9**:1572

[29] Bhattacharya A. Effect of soil water deficit on growth and development of plants: A review. In: Soil Water

Deficit and Physiological Issues in Plants. 2021. pp. 393-488. Available from: https://link.springer.com/chapter/10.1007/978-981-33-6276-5_5

[30] Zhu M, He Y, Zhu M, Ahmad A, Xu S, He Z, et al. ipa1 improves rice drought tolerance at seedling stage mainly through activating abscisic acid pathway. Plant Cell Reports. 2022;**41**(1):221-232

[31] Wu J, Wang J, Hui W, Zhao F, Wang P, Su C, et al. Physiology of plant responses to water stress and related genes: A review. Forests. 2022;**13**(2):324

[32] Zhao W, Liu L, Shen Q, Yang J, Han X, Tian F, et al. Effects of water stress on photosynthesis, yield, and water use efficiency in winter wheat. Water. 2020;**12**(8):2127

[33] Zandi P, Schnug E. Reactive oxygen species, antioxidant responses and implications from a microbial modulation perspective. Biology. 2022;**11**(2):155

[34] Kapoor D, Bhardwaj S, Landi M, Sharma A, Ramakrishnan M, Sharma A. The impact of drought in plant metabolism: How to exploit tolerance mechanisms to increase crop production. Applied Sciences. 2020;**10**(16):5692

[35] MacLean A, Legendre F, Appanna VD. The tricarboxylic acid (TCA) cycle: A malleable metabolic network to counter cellular stress. Critical Reviews in Biochemistry and Molecular Biology. 2023;**58**(1):81-97

[36] Khan N, Ali S, Zandi P, Mehmood A, Ullah S, Ikram M, et al. Role of sugars, amino acids and organic acids in improving plant abiotic stress tolerance. Pakistan Journal of Botany. 2020;**52**(2):355-363

[37] Jiang S, Lan Z, Zhang Y, Kang X, Zhao L, Wu X, et al. Mechanisms by which exogenous substances enhance plant salt tolerance through the modulation of ion membrane transport and reactive oxygen species metabolism. Antioxidants. 2024;**13**(9):1050

[38] Akram NA, Fatima K, Kong H, Zafar N, Mahmood S, Ashraf M, et al. Interactive effect of drought stress and l-methionine on the growth and physio-biochemical changes in broccoli (*Brassica oleracea* L. var. italica): Leaf and head. Journal of Plant Growth Regulation. 2024;**43**(6):1954-1966

[39] Xie H, Bai G, Lu P, Li H, Fei M, Xiao BG, et al. Exogenous citric acid enhances drought tolerance in tobacco (*Nicotiana tabacum*). Plant Biology. 2022;**24**(2):333-343

[40] Chatterjee A. Quantitative genetics. In: Genetics Fundamentals Notes. Singapore: Springer; 2022. pp. 1029-1076

[41] Singh C, Yadav S, Khare V, Gupta V, Patial M, Kumar S, et al. Wheat drought tolerance: Unveiling a synergistic future with conventional and molecular breeding strategies. Plants. 2025;**14**(7):1053

[42] Rajendra V. Marker-assisted selection (MAS) revolutionized: Integrating 'omics' data into tomato breeding. In: Omics Approaches for Tomato Yield and Quality Trait Improvement. Cham: Springer; 2025. pp. 273-305

[43] Mahmood U, Li X, Fan Y, Chang W, Niu Y, Li J, et al. Multi-omics revolution to promote plant breeding efficiency. Frontiers in Plant Science. 2022;**13**:1062952

[44] Khondoker M, Mandal S, Gurav R, Hwang S. Freshwater shortage, salinity increase, and global food production:

A need for sustainable irrigation water desalination—A scoping review. Earth. 2023;**4**(2):223-240

[45] Demo AH, Gemeda MK, Abdo DR, Guluma TN, Adugna DB. Impact of soil salinity, sodicity, and irrigation water salinity on crop production and coping mechanism in areas of dryland farming. Agrosystems, Geosciences & Environment. 2025;**8**(1):e70072

[46] Ulas F, Ameen HHHH, Ulas A, Yücel YC. Mitigating the effect of salt stress through grafting technology in vegetables: A review. Journal of Crop Health. 2025;**77**(4):1-21

[47] Khan AA, Wang Y-F, Akbar R, Alhoqail WA. Mechanistic insights and future perspectives of drought stress management in staple crops. Frontiers in Plant Science. 2025;**16**:1547452

[48] Jiang Z, van Zanten M, Sasidharan R. Mechanisms of plant acclimation to multiple abiotic stresses. Communications Biology. 2025;**8**(1):655

[49] Nauman M, Iqbal N, Parveen N, Jamil HMA, Riyazuddin R. Anatomical, morphological, and physiological consequences and adaptations in flood-stressed plants. In: Plant Flooding: Sensitivity and Tolerance Mechanisms. Cham: Springer; 2025. pp. 3-27

[50] Hmissi M, Chaieb M, Krouma A. Exogenous IAA application enhances durum wheat tolerance to salinity by regulating osmotic adjustment and ionic homeostasis. Russian Journal of Plant Physiology. 2025;**72**(3):60

[51] Dharmapriya T, Rathnayake I, Deeyamulla M, Liyandeniya A, Madhushani L, Priyantha N. Comparison of Ionic Concentrations (Na+, Ca2+, Mg2+, NH4+, Cl-, SO42-, NO3-) and Water Quality Parameters of Bulk Depositions among Three Sites in Kandy Area. Sri Lanka; 2023. Available from: https://www.researchsquare.com/article/rs-2731158/v1

[52] Grigore M-N. Definition and classification of halophytes as an ecological group of plants. In: Handbook of Halophytes: From Molecules to Ecosystems towards Biosaline Agriculture. Cham: Springer; 2021. pp. 3-50

[53] Rahman MM, Mostofa MG, Keya SS, Siddiqui MN, Ansary MMU, Das AK, et al. Adaptive mechanisms of halophytes and their potential in improving salinity tolerance in plants. International Journal of Molecular Sciences. 2021;**22**(19):10733

[54] Farooq M, Rafique S, Zahra N, Rehman A, Siddique KH. Root system architecture and salt stress responses in cereal crops. Journal of Agronomy and Crop Science. 2024;**210**(6):e12776

[55] Cárdenas-Pérez S, Niedojadło K, Mierek-Adamska A, Dąbrowska G, Piernik A. Maternal salinity influences anatomical parameters, pectin content, biochemical and genetic modifications of two *Salicornia europaea* populations under salt stress. Scientific Reports. 2022;**12**(1):2968

[56] Arif Y, Singh P, Siddiqui H, Bajguz A, Hayat S. Salinity induced physiological and biochemical changes in plants: An omic approach towards salt stress tolerance. Plant Physiology and Biochemistry. 2020;**156**:64-77

[57] Okon OG. Effect of salinity on physiological processes in plants. In: Microorganisms in Saline Environments: Strategies and Functions. Springer; 2019. pp. 237-262

[58] Stefanov MA, Rashkov GD, Yotsova EK, Dobrikova AG, Apostolova EL. Impact of salinity on the

energy transfer between pigment–protein complexes in photosynthetic apparatus, functions of the oxygen-evolving complex and photochemical activities of photosystem II and photosystem I in two paulownia lines. International Journal of Molecular Sciences. 2023;**24**(4):3108

[59] Hasanuzzaman M, Raihan MRH, Masud AAC, Rahman K, Nowroz F, Rahman M, et al. Regulation of reactive oxygen species and antioxidant defense in plants under salinity. International Journal of Molecular Sciences. 2021;**22**(17):9326

[60] AL-Aloosy YAM, AL-Tameemi AJ, Jumaa SS. The role of enzymatic and non-enzymatic antioxidants in facing the environmental stresses on plant: A review. Plant Archives. 2019;**19**(1):1057-1060

[61] Fujita M, Hasanuzzaman M. Approaches to Enhancing Antioxidant Defense in Plants. Vol. 11. Switzerland: MDPI; 2022

[62] Shahrajabian MH, Cheng Q, Sun W. The most important medicinal herbs and plants in traditional Chinese and Iranian medicinal sciences with antioxidant activities. Letters in Drug Design & Discovery. 2023;**20**(9):1171-1184

[63] Kumari A, Sharma B, Singh BN, Hidangmayum A, Jatav HS, Chandra K, et al. Physiological mechanisms and adaptation strategies of plants under nutrient deficiency and toxicity conditions. In: Plant Perspectives to Global Climate Changes. Elsevier; 2022. pp. 173-194. Available from: https://www.sciencedirect.com/science/chapter/edited-volume/abs/pii/B9780323856652000121?via%3Dihub

[64] Sardans J, Peñuelas J. Potassium control of plant functions: Ecological and agricultural implications. Plants. 2021;**10**(2):419

[65] Dubey RS, Srivastava RK, Pessarakli M. Physiological mechanisms of nitrogen absorption and assimilation in plants under stressful conditions. In: Handbook of Plant and Crop Physiology. New York, NY: CRC Press; 2021. pp. 579-616

[66] Alsaeedi A, El-Ramady H, Alshaal T, El-Garawany M, Elhawat N, Al-Otaibi A. Silica nanoparticles boost growth and productivity of cucumber under water deficit and salinity stresses by balancing nutrients uptake. Plant Physiology and Biochemistry. 2019;**139**:1-10

[67] Chaudhary N, Salgotra RK, Chauhan BS. Genetic enhancement of cereals using genomic resources for nutritional food security. Genes. 2023;**14**(9):1770

[68] Singh RK, Kota S, Flowers TJ. Salt tolerance in rice: Seedling and reproductive stage QTL mapping come of age. Theoretical and Applied Genetics. 2021;**134**(11):3495-3533

[69] Ullah MA, Abdullah-Zawawi M-R, Zainal-Abidin R-A, Sukiran NL, Uddin MI, Zainal Z. A review of integrative omic approaches for understanding rice salt response mechanisms. Plants. 2022;**11**(11):1430

[70] Patra S, Chatterjee D, Basak S, Sen S, Mandal A. CRISPR/Cas9 opens new horizon of crop improvement under stress condition. Biochimica et Biophysica Acta (BBA)-General Subjects. 2024;**1868**(10):130685

[71] Fernando VD. Major transcription factor families involved in salinity stress tolerance in plants. In: Transcription Factors for Abiotic Stress Tolerance in Plants. 2020. pp. 99-109. Available from: https://www.sciencedirect.com/science/chapter/edited-volume/abs/pii/B9780128193341000071?via%3Dihub

[72] Lu B, Dao PD, Liu J, He Y, Shang J. Recent advances of hyperspectral imaging technology and applications in agriculture. Remote Sensing. 2020;**12**(16):2659

[73] Roychowdhury R, Das SP, Gupta A, Parihar P, Chandrasekhar K, Sarker U, et al. Multi-omics pipeline and omics-integration approach to decipher plant's abiotic stress tolerance responses. Genes. 2023;**14**(6):1281

[74] Haque E, Taniguchi H, Hassan MM, Bhowmik P, Karim MR, Śmiech M, et al. Application of CRISPR/Cas9 genome editing technology for the improvement of crops cultivated in tropical climates: Recent progress, prospects, and challenges. Frontiers in Plant Science. 2018;**9**:617

[75] Kamali S, Singh A. Genomic and transcriptomic approaches to developing abiotic stress-resilient crops. Agronomy. 2023;**13**(12):2903

[76] Kumar P, Paul D, Jhajhriya S, Kumar R, Dutta S, Siwach P, et al. Understanding heat-shock proteins' abundance and pivotal function under multiple abiotic stresses. Journal of Plant Biochemistry and Biotechnology. 2024;**33**(4):492-513

[77] Kumar M, Kumar Patel M, Kumar N, Bajpai AB, Siddique KH. Metabolomics and molecular approaches reveal drought stress tolerance in plants. International Journal of Molecular Sciences. 2021;**22**(17):9108

[78] Naz M, Ghani MI, Atif MJ, Raza MA, Bouzroud S, Afzal MR, et al. Sodium and abiotic stress tolerance in plants. In: Beneficial Chemical Elements of Plants: Recent Developments and Future Prospects. 2023. pp. 307-330. Available from: https://onlinelibrary.wiley.com/doi/10.1002/9781119691419.ch13

[79] Song L, Wang R, Yang X, Zhang A, Liu D. Molecular markers and their applications in marker-assisted selection (MAS) in bread wheat (*Triticum aestivum* L.). Agriculture. 2023;**13**(3):642

[80] Alex S, Augustine R. CRISPR/Cas9-mediated genome editing for abiotic stress tolerance in crops: Current advances and future prospects. Plant Molecular Biology Reporter. 2025;**43**(4):1767-1789

[81] Anwar A, Kim J-K. Transgenic breeding approaches for improving abiotic stress tolerance: Recent progress and future perspectives. International Journal of Molecular Sciences. 2020;**21**(8):2695

[82] Rajawat J. Transcriptomics. In: Omics Approaches, Technologies and Applications: Integrative Approaches for Understanding OMICS Data. Cham: Springer; 2019. pp. 39-56

[83] Kaya C. Molecular crosstalk of transcription factors and hormones in drought tolerance genes in plants. Turkish Journal of Botany. 2025;**49**(3):145-164

[84] Liu Y, Lu S, Liu K, Wang S, Huang L, Guo L. Proteomics: A powerful tool to study plant responses to biotic stress. Plant Methods. 2019;**15**(1):135

[85] Imran H, Gul A, Paracha RZ, Amir R, Munir F, Bhatti MF. In silico identification and functional annotation of universal stress protein (USP) gene family in *Chenopodium quinoa*. Scientific Reports. 2025;**15**(1):18264

[86] Muthubharathi BC, Gowripriya T, Balamurugan K. Metabolomics: Small molecules that matter more. Molecular Omics. 2021;**17**(2):210-229

[87] Xu Y, Fu X. Reprogramming of plant central metabolism in response to

abiotic stresses: A metabolomics view. International Journal of Molecular Sciences. 2022;**23**(10):5716

[88] Ali S, Tyagi A, Bae H. Ionomic approaches for discovery of novel stress-resilient genes in plants. International Journal of Molecular Sciences. 2021;**22**(13):7182

[89] Malakar P, Chattopadhyay D. Adaptation of plants to salt stress: The role of the ion transporters. Journal of Plant Biochemistry and Biotechnology. 2021;**30**(4):668-683

[90] Landi M, Misra BB, Nocito FF, Lucchini G, Bruno L, Malara A, et al. Metabolic changes induced by *Cuscuta campestris* Yunck in the host species *Artemisia campestris* subsp. *variabilis* (ten.) Greuter as a strategy for successful parasitisation. Planta. 2022;**256**(6):118

[91] Sahito JH, Zhang H, Gishkori ZGN, Ma C, Wang Z, Ding D, et al. Advancements and prospects of genome-wide association studies (GWAS) in maize. International Journal of Molecular Sciences. 2024;**25**(3):1918

[92] Baloch FS, Altaf MT, Liaqat W, Bedir M, Nadeem MA, Cömertpay G, et al. Recent advancements in the breeding of sorghum crop: Current status and future strategies for marker-assisted breeding. Frontiers in Genetics. 2023;**14**:1150616

[93] Li Y, Li W, Li J. The CRISPR/Cas9 revolution continues: From base editing to prime editing in plant science. Journal of Genetics and Genomics. 2021;**48**(8):661-670

[94] Shelake RM, Kadam US, Kumar R, Pramanik D, Singh AK, Kim J-Y. Engineering drought and salinity tolerance traits in crops through CRISPR-mediated genome editing: Targets, tools, challenges, and perspectives. Plant Communications. 14 Nov 2022;**3**(6):100417. DOI: 10.1016/j.xplc.2022.100417

[95] Gao C. Genome engineering for crop improvement and future agriculture. Cell. 2021;**184**(6):1621-1635

[96] Şimşek Ö, Isak MA, Dönmez D, Dalda Şekerci A, İzgü T, Kaçar YA. Advanced biotechnological interventions in mitigating drought stress in plants. Plants. 2024;**13**(5):717

[97] Dar MH, Bano DA, Waza SA, Zaidi NW, Majid A, Shikari AB, et al. Abiotic stress tolerance-progress and pathways of sustainable rice production. Sustainability. 2021;**13**(4):2078

[98] Singer SD, Laurie JD, Bilichak A, Kumar S, Singh J. Genetic variation and unintended risk in the context of old and new breeding techniques. Critical Reviews in Plant Sciences. 2021;**40**(1):68-108

[99] Tripathi D, Singh M, Pandey-Rai S. Crosstalk of nanoparticles and phytohormones regulate plant growth and metabolism under abiotic and biotic stress. Plant Stress. 2022;**6**:100107

[100] Li W, Li W-B, Xing L-J, Guo S-X. Effect of arbuscular mycorrhizal fungi (AMF) and plant growth-promoting rhizobacteria (PGPR) on microorganism of phenanthrene and pyrene contaminated soils. International Journal of Phytoremediation. 2023;**25**(2):240-251

[101] Shoukat A, Pitann B, Zafar MM, Farooq MA, Haroon M, Nawaz A, et al. Nanotechnology for climate change mitigation: Enhancing plant resilience under stress environments. Journal of Plant Nutrition and Soil Science. 2024;**187**(5):604-620

[102] Hossain A, Skalicky M, Brestic M, Maitra S, Ashraful Alam M, Syed MA, et al. Consequences and mitigation strategies of abiotic stresses in wheat (*Triticum aestivum* L.) under the changing climate. Agronomy. 2021;**11**(2):241

[103] Rosero A, Granda L, Berdugo-Cely JA, Šamajová O, Šamaj J, Cerkal R. A dual strategy of breeding for drought tolerance and introducing drought-tolerant, underutilized crops into production systems to enhance their resilience to water deficiency. Plants. 2020;**9**(10):1263

[104] Sackey OK, Feng N, Mohammed YZ, Dzou CF, Zheng D, Zhao L, et al. A comprehensive review on rice responses and tolerance to salt stress. Frontiers in Plant Science. 2025;**16**:1561280

[105] Kumari K, Mahato M, Krishna VJ, Bhaduri D. New-generation mitigation practices for drought Management in Rice. In: Drought and Heat Stress in Agriculture: Implications, Mitigation and Policy Approaches. Cham: Springer; 2025. pp. 329-364

[106] Tamilarasi M. Molecular Characterization of Salt Tolerance in Wheat Var. Kharchia 65. Division of Molecular Biology and Biotechnology ICAR-Indian Agricultural; 2020. Available from: https://krishikosh.egranth.ac.in/items/aa9be045-60a0-4cb7-8420-b4d1b09a0a2a

[107] Sharma I, Singh G, Gupta R. Wheat improvement in India. In: Improving Wheat Productivity in Asia. 2013. p. 81. Available from: https://www.apaari.org/wp-content/uploads/downloads/2013/05/Improving-Wheat-Productivity-in-Asia-2013.pdf

[108] Badu-Apraku B, Fakorede MA, Nelimor C, Osuman AS, Bonkoungou TO, Muhyideen O, et al. Recent advances in breeding maize for drought, heat and combined heat and drought stress tolerance in sub-Saharan Africa. CABI Reviews. 2023. Available from: https://www.cabidigitallibrary.org/doi/full/10.1079/cabireviews.2023.0011

Chapter 3

Phytohormonal Regulation of Photosynthesis under Optimal and Abiotic Stress Conditions

Abu Bakar Siddique, Ayesha Siddiqua, Mst. Mahzabin Zannat and Mohammad Shah Jahan

Abstract

Photosynthesis is an essential physiological process in plants, facilitating their growth and development by transforming light energy into chemical energy. Phytohormones, as endogenous signaling chemical compounds, profoundly impact this process. The main phytohormones—auxins, gibberellins, cytokinins, abscisic acid, and ethylene—are important for the photosynthetic pathway. Phytohormonal regulation encompasses all aspects of the photosynthetic process, including stomatal regulation and gas exchange, maintenance of the photosynthetic apparatus, biosynthesis of key photosynthetic enzymes such as Rubisco, chlorophyll biosynthesis, and the expression of major genes involved in photosynthesis. Beyond their roles under optimal conditions, phytohormones exhibit both synergistic and antagonistic effects on photosynthesis under abiotic stress. In both scenarios, hormonal crosstalk occurs in a highly complex manner. This chapter provides an overview of the significant roles of phytohormones in photosynthesis and elucidates the regulatory networks linking phytohormone signaling to photosynthetic efficiency through biochemical and molecular mechanisms. Additionally, it addresses the interactions and crosstalk among hormones that influence photosynthesis under abiotic stress conditions. Finally, the chapter highlights future research directions aimed at expanding our understanding of phytohormonal regulation of photosynthesis and its potential application in stress mitigation strategies.

Keywords: phytohormones, stomatal regulation, chlorophyll synthesis, carbon assimilation, climate change, hormonal crosstalk

1. Introduction

Photosynthesis is a biochemical process utilized by plants, algae, and certain bacteria to transform light energy into chemical energy, which is subsequently stored in the form of carbohydrates [1]. This process may also be characterized as the synthesis of organic compounds from inorganic substances and solar energy within plant systems. Serving as the cornerstone of life on Earth, photosynthesis facilitates the energy requirements essential for plant development and, by extension, supports the entire food web. In the context of agriculture, the effectiveness of photosynthesis significantly influences crop yield, thereby impacting global food production [2]. This vital process occurs in all green plants—both lower and higher forms—as well as in photosynthetic bacteria [3]. Chloroplasts, which house chlorophyll, represent the primary sites of photosynthesis and play a crucial role in the growth and development of plants. While leaves are predominantly the sites of photosynthesis, they are not the sole tissues within plants where this process occurs. In addition to leaves, several other organs, such as petioles, stem, fruits, and ears and husks of some plant (wheat), have been reported to exhibit this process [4]. Photosynthesis is a fundamental process essential for plant life, converting sunlight, carbon dioxide, and water into glucose for plant growth and oxygen for sustaining life, and as a primary driver of plant productivity, it plays a crucial role in agriculture through two interrelated stages. The first step is light-dependence while the second step is light-independence cycles [5].

Photosynthesis is complex process involving several components and physiological processes which are affected by both endogenous and exogenous factors [6]. One of the major endogenous regulators affecting photosynthesis is the endogenous level of phytochrome. Plant hormones, also referred to as plant growth regulators (PGRs), are chemical substances found in plants at minimal concentrations. Phytohormones—including auxins, cytokinins, gibberellins, abscisic acid (ABA), ethylene, jasmonates, salicylic acid, and brassinosteroids—play a significant role in modulating photosynthetic efficiency. Some of them have positive impact on photosynthesis processes while some others show negative regulation on key events in photosynthesis process [7]. One of the direct effects of phytohormones involves the biosynthesis of chlorophyll and the development of chloroplast in plants [8, 9]. Regulation of stomatal movement (closure and opening) has a direct effect on photosynthesis process, including CO_2 intake, assimilation and gas exchange [10, 11]. By regulating the stomatal aperture, stomatal number, stomatal density, and stomatal distribution, phytohormones affect the photosynthesis rate [12–14]. In addition to regulating the anatomical features, phytohormones are involved at physiological and molecular level such as enzymatic activity and gene expression, affecting the photosynthesis process [7, 15]. In addition to their roles under optimum conditions, phytohormones are involved in regulating the photosynthesis process under abiotic stress conditions, either synergistically or antagonistically among them [16], by affecting the stomatal movement, chlorophyll biosynthesis, and protecting photosynthetic apparatus. Photosynthesis is essential for plant growth and supports the global food web. Photosynthesis occurs in various plant tissues, including leaves, petioles, stems, fruits, and even parts of wheat like ears and husks. Phytohormones play a significant role in modulating photosynthetic efficiency. They affect chlorophyll biosynthesis, chloroplast development, and stomatal movement, which directly influence CO_2 intake and gas exchange. Phytohormones also impact photosynthesis at the physiological and molecular levels, including enzymatic activity and gene expression. Under abiotic stress conditions, phytohormones regulate photosynthesis through various mechanisms, either synergistically or antagonistically. In this chapter, we revisit the phytohormonal roles in different events and features of plant photosynthesis. In addition to discussing their single effect, we also provide insight into their crosstalk in regulating the photosynthesis under both optimum and abiotic stress conditions. Finally, this chapter has outlines some of the key research areas that can be targeted for future study.

2. Phytohormones and photosynthesis process

Plant hormones have direct and indirect control on diverse aspects of plant photosynthesis, including chloroplast development, stomatal regulations, and photoprotection. These processes are associated with photosynthetic rates and efficiency under both optimum and diverse stress conditions. The following subsections discuss the roles of individual phytohormones in different steps of photosynthesis process in plants and are visualized in **Figure 1**.

2.1 Auxin and photosynthesis

Auxin plays diverse roles in all aspects of plant growth and development, which has been well documented [17]. Auxin contributes to several aspects of photosynthesis, including structural development, and leaf surface area expansion, which enhances the light harvesting efficiency and thereby increases the photosynthesis under optimum conditions. Leaf architecture has direct impact on photosynthesis process. Auxin has been reported to play crucial roles in chloroplast development, biosynthesis of chlorophyll content, and modulation of photosynthetic enzyme activity [18]. Leaf expansion has an indirect effect on photosynthetic efficiency as it enhances the light harvesting capacity [19]. Furthermore, auxin signaling is essential for the proper development and functionality of chloroplasts, as it affects the expression of genes involved in photosynthesis [18]. For example, the external application of IAA at a concentration of 10 μM markedly improved both the photosynthetic rate and chlorophyll content in Arabidopsis. However, elevated levels of auxin (>50 μM) can disturb hormonal equilibrium, resulting in abnormal

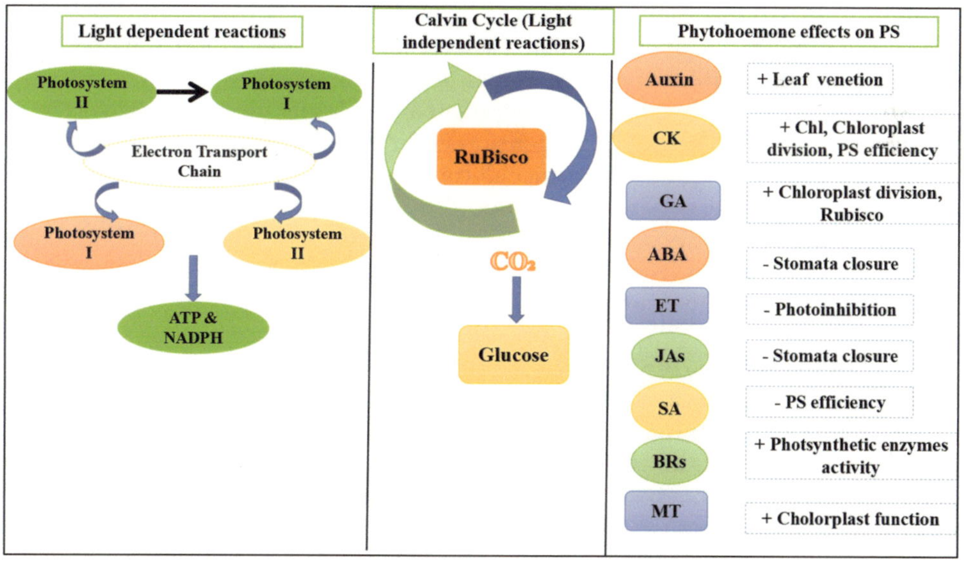

Figure 1.
Phytohormonal roles on plant photosynthesis process. Phytohormonal regulations involve chlorophyll biosynthesis, stomatal regulations, and leaf venation. ATP- adenosine triphosphate; CK-cytokinin; GA-gibberellic acid; ABA-abscisic acid; ET-ethylene; JAs-jasmonic acids; SA-salicylic acid; BRs-brassinosteroids; MT-melatonin; Chl-chlorophyll; PS-photosystem; CO_2-carbondi oxside; NADPH-nicotinamide adenine dinucleotide phosphate; + denotes the positive effects and - denotes the negative effects.

leaf structure and a reduction in the photosynthetic area [20, 21]. Excess auxin can suppress chloroplast development, resulting in decreased chlorophyll synthesis and impaired photosynthesis. Another aspect of auxin roles in photosynthesis is by regulating Rubisco activity. Rubisco is one of the central enzymes that regulate the photosynthesis process which is upregulated by auxin [22]. However, auxins' regulation on photosynthesis is mainly dictated by its concentration as over accumulation of auxin leads to increased reactive oxygen species production, leading to degradation of photosynthetic proteins [23]. Additionally, excess auxin level results in stomatal closure and limits the CO_2 uptake and reduces photosynthetic efficiency [24].

2.2 Cytokinins and photosynthesis

Cytokinins are vital plant hormones that play a key role in regulating and supporting photosynthesis. They promote chloroplast development, enhance chlorophyll biosynthesis, upregulate essential photosynthetic components such as Rubisco, and improve the plant's ability to withstand stress. Although cytokinins are primarily known for facilitating cell division and delaying senescence, they also exert both direct and indirect effects on photosynthetic efficiency. These effects are largely attributed to their influence on chloroplast development and the biosynthesis of photosynthetic pigments through the upregulation of genes involved in these processes [7]. Cytokinins (CKs) delay leaf senescence, thereby extending the duration of photosynthetic activity. They regulate a broad range of photosynthesis-related genes and help maintain the integrity and efficiency of the photosynthetic system, particularly under stress conditions [25]. Moreover, CKs and their derivatives exert protective effects on the photosynthetic apparatus. These include the upregulation of photosystem II (and occasionally photosystem I) components, enhanced expression of Calvin cycle enzymes such as Rubisco and glyceraldehyde-3-phosphate dehydrogenase, and stabilization of chlorophyll levels through the suppression of its degradation [26]. Since their initial discovery, cytokinins have been consistently recognized for their central role in chloroplast development and chlorophyll synthesis [8]. They are known to regulate more than 100 genes associated with photosynthesis, including those encoding the light-harvesting chlorophyll a/b-binding proteins of PSII and both the small and large subunits of Rubisco (RBCS and RBCL), which are crucial for carbon fixation [25].

2.3 Gibberellins and photosynthesis

Gibberellic acid (GA_3) is a vital plant hormone well-known for its dynamic effects on promoting stem growth, triggering seed germination, and enhancing flowering, which make it one of the essential phytohormones for normal growth [27]. Nevertheless, its effects also encompass photosynthesis, as it alters leaf expansion, chloroplast development, and the activity of various enzymes. While GA_3 does not directly facilitate photosynthesis in the manner that light or carbon dioxide do, it establishes beneficial physiological conditions that improve the efficiency of photosynthetic processes. One of the major roles is the effect on chloroplast development. GA plays a positive role in enhancing photosynthesis activity and promotes chloroplast biogenesis [28]. Increasing the leaf area enhances the light absorption process. GA_3 promotes cell elongation and division, resulting in larger leaves with greater surface area for light absorption. GA_3 enhances chlorophyll biosynthesis and improves chloroplast structure, boosting light harvesting. For example, exogenous application of GA_3 increased chlorophyll a and b content in wheat, associated with a higher photosynthesis rate [29]. GA_3 stimulates the activity of Rubisco and other Calvin cycle enzymes, improving carbon fixation and photosynthetic enzyme activation. In one study, exogenous application of GA_3 treatment increased Rubisco activity, resulting higher net photosynthesis in rice [30]. However, GA_3 has also a harmful effect on photosynthesis because a higher GA_3 level can trigger chlorophyll degradation, especially under stress conditions, resulting in reduced chlorophyll content and lower photosynthesis as observed in soybean [31].

2.4 Abscisic acid (ABA) and photosynthesis

Abscisic acid (ABA), a lipid-derived phytohormone, generally exerts a negative influence on photosynthesis by inducing stomatal closure under stress conditions. This closure restricts the diffusion of CO_2 into the leaf, thereby reducing the photosynthetic rate. In addition to limiting CO_2 availability, ABA can interfere with key photosynthetic enzymes such as Rubisco and disrupt cellular processes, including electron transport, ultimately diminishing the plant's photosynthetic efficiency. Despite its inhibitory effects, ABA also plays a crucial protective role during abiotic stress. By regulating stomatal closure, it minimizes transpirational water loss and activates antioxidant defense mechanisms, helping plants sustain photosynthetic activity under drought or salinity conditions. ABA contributes to maintaining the structural and functional integrity of photosystem II (PSII) and Rubisco under salinity stress, thereby preserving photosynthetic efficiency [32]. However, excessive accumulation of ABA, particularly under prolonged stress or during leaf senescence, can suppress chlorophyll biosynthesis and accelerate the decline of photosynthetic pigments [23].

2.5 Ethylene and photosynthesis

Ethylene, a gaseous phytohormone, plays a bifunctional role in regulating plant photosynthesis. While basal levels of ethylene are essential for the normal functioning of the photosynthetic machinery, elevated levels—often induced by environmental stress—can cause degradation of the photosynthetic apparatus and chlorophyll in mature leaves, thereby accelerating the senescence process. In addition, ethylene influences the regulation of chlorophyll content in plants [33]. Ethylene affects chloroplast development, chlorophyll accumulation, and overall photosynthetic performance through multiple mechanisms that vary depending on plant species, developmental stage, light conditions, and environmental stress factors [9]. The regulation of stomatal responses by ethylene is particularly complex, often exhibiting dual or even contradictory effects depending on the physiological and environmental context [34], which necessitate the further experimental-based evidence to understand the ethylene-induced stomatal regulations.

2.6 Brassinosteroids (BRs) and photosynthesis

Brassinosteroids (BRs), a distinct class of plant hormones, are naturally occurring polyhydroxylated sterol derivatives found in all plant species. In young tissues, BR concentrations are considerably higher than in mature tissues, typically ranging from 1 to 100 ng/g fresh weight compared to 0.01 to 0.1 ng/g in mature tissues [35].

BRs enhance photosynthesis by increasing chlorophyll content, stabilizing photosystem II (PSII), facilitating stomatal conductance for efficient carbon dioxide uptake, and activating enzymes involved in the Calvin cycle. They protect photosynthetic pigments by stimulating the synthesis of enzymes responsible for chlorophyll biosynthesis and maintaining PSII efficiency through the stabilization of the D1 protein. By alleviating stomatal limitations, BRs improve carbon fixation efficiency, while their involvement in both light and dark reactions contributes to greater carbohydrate production. Studies have demonstrated that BRs safeguard chloroplast structure and enhance CO_2 absorption and utilization by increasing the activity of key components such as the cytochrome b_6/f complex, D1 and D2 proteins, Rubisco's large subunit, and carbonic anhydrase, all of which collectively regulate the overall photosynthetic rate [11, 36]. Furthermore, BRs enhance photosynthetic efficiency by upregulating gene expression associated with PSII photochemical performance, chlorophyll biosynthesis, and the activity of critical Calvin cycle enzymes. They also improve photosystem performance and stimulate NADPH production, a key energy carrier in photosynthesis [11]. Enhanced CO_2 assimilation has similarly been observed in plants with elevated BR levels following exogenous application [37, 38]. In summary, BRs serve as vital regulators of photosynthetic efficiency, acting at multiple molecular and physiological levels to optimize light capture, carbon fixation, and energy production, thereby supporting plant growth and stress resilience.

2.7 Salicylic acid (SA) and photosynthesis

Salicylic acid (SA) is a versatile plant hormone that can exert both beneficial and detrimental effects on photosynthesis, depending on factors such as its mode of application, concentration, plant species, and prevailing environmental conditions. As a key regulator of photosynthetic processes, SA significantly influences chlorophyll content, stomatal conductance, and the activity of photosynthesis-related enzymes [13]. It is widely recognized for enhancing photosynthetic efficiency and improving the overall function of the photosynthetic apparatus. The application of exogenous SA has been shown to increase pigment concentrations and enhance photosynthetic performance in tomato plants [39]. The response of plants to SA treatment is closely linked to the modulation of enzymatic activities involved in photosynthesis. SA functions as a stress-related signaling molecule that activates abiotic stress-responsive genes [40]. Notably, SA applications have been associated with increased production of heat shock proteins, molecular chaperones, and dehydrin-like proteins, all of which contribute to stress tolerance. In addition, changes in Rubisco activity, chlorophyll concentration, and protein kinase activity have been reported following SA treatment [41, 42]. In summary, salicylic acid plays a multifaceted role in regulating photosynthesis by modulating chlorophyll biosynthesis, enzyme activity, and stress-responsive signaling pathways, ultimately contributing to enhanced photosynthetic performance and stress adaptation in plants.

2.8 Jasmonic acid (JA) and photosynthesis

Jasmonic acid (JA) and its derivatives—such as jasmonoyl isoleucine (JA-Ile), methyl jasmonate (MeJA), and other oxylipins—are collectively known as jasmonates [43]. One of the earliest documented physiological effects of jasmonates is their role in promoting leaf senescence, which is often associated with reduced chlorophyll content and the downregulation of photosynthesis-related genes. JA significantly influences the expression of genes involved in photosynthetic regulation, enabling plants to better tolerate adverse conditions such as drought and metal toxicity. Because JA is synthesized in the chloroplast, the primary site of photosynthesis, it maintains a close functional relationship with the regulation and protection of the photosynthetic apparatus. The methyl ester of jasmonic acid (MeJA) has been shown to selectively inhibit chlorophyll accumulation at the precursor stage in the dark, distinguishing it from other growth regulators under varying environmental conditions [44]. The inhibitory effects of JA are largely attributed to its epinastic influence, which suppresses leaf expansion and chlorophyll accumulation. Furthermore, JA can induce chlorophyll degradation and disrupt microtubule structures in certain crop species [45]. MeJA, in particular, is well recognized for its inhibitory role in stomatal opening, photosynthetic activity, and overall plant growth [46]. Interestingly, not all effects of JA on photosynthesis are negative. Some studies have reported that JA promotes chlorophyll accumulation, potentially by enhancing nutrient and water uptake [47]. Exogenous JA application has also been linked to improved magnesium absorption—a crucial element for chlorophyll biosynthesis—thereby supporting photosynthetic activity under drought stress [48]. In conclusion, jasmonic acid exerts a dual role in the regulation of photosynthesis, acting as both an inhibitor and a protector depending on its concentration, plant developmental stage, and environmental conditions, ultimately balancing growth regulation and stress adaptation.

Phytohormones	Biosynthesis location	Effect on photosynthesis	References
Auxin	Young leaf and shoot tip	Expansion of leaf surface area and enhancement of light harvesting process	Khan et al. [51]
Cytokinins	Primarily in roots	Photosynthetic enzymatic activity and chloroplast development	Nagar et al. [52]
Gibberellin	Young leaf, growing stem, and sprouted seed	Higher light perception and photosynthetic gene upregulation	Razzaq and Du [53]
Abscisic Acid (ABA)	Leaf and root, mainly under stress	Stomatal regulation, CO_2 intake and assimilation	Bagal et al. [54]
Ethylene	All organs	Reduces photosynthetic capacity by accelerating senescence	Mohorović et al. [55]
Jasmonic Acid	Leaf and tissues upon wounding mainly	Reduce photosynthesis by suppressing chloroplasts activity during stress condition	Wasternack [56]
Salicylic acid	Leaf and shoot upon biotic and abiotic stresses	Suppress photosynthesis during pathogen attack	Cheaib and Killiny [57]
Brassinosteroids	Active growing organs	Enhanced photosynthetic enzymes activity and chloroplast's function	Nolan et al. [58]
Melation	leaf and root, mainly under stress.	Maintain photosynthesis by protecting chloroplasts from oxidation	Hussain et al. [59]

Table 1.
Phytohormonal regulations on photosynthesis process in plants.

2.9 Melatonin and photosynthesis

Melatonin acts as a versatile signaling molecule that regulates plant growth and stress responses. It positively influences photosynthesis by protecting light-harvesting pigments, enhancing the efficiency of photosystems responsible for converting light into chemical energy, and modulating the expression of genes involved in photosynthetic activity [49]. Recent studies have shown that seed treatment with melatonin can improve photosynthetic efficiency in plants [15]. In sugar beet subjected to drought stress, melatonin significantly increased stomatal conductance, facilitated leaf gas exchange, elevated chlorophyll content, and supported overall photosynthetic performance. Additionally, under low-temperature stress, melatonin enhanced photosynthetic efficiency by promoting the activity of key carbon fixation enzymes, including ribulose-1,5-bisphosphate carboxylase/oxygenase (Rubisco) [50]. In summary, melatonin acts as a critical regulator of photosynthesis, enhancing pigment protection, enzyme activity, and gas exchange, thereby improving photosynthetic performance and stress resilience in plants.

Based on the above-discussion, **Table 1** summarizes the direct and indirect roles of different phytohormones on photosynthesis process in plants.

3. Crosstalk between phytohormones and photosynthesis

Phytohormones, in different combinations, are involved in complex ways to regulate various aspects of photosynthesis, including chloroplast biogenesis, chlorophyll biosynthesis and degradation, stomatal regulation and gas exchange, photosynthetic enzymatic activity, light-harvesting processes, and leaf development [60]. These events directly or indirectly regulate photosynthesis in plants. Many of these processes are controlled by the interplay of more than one or two phytohormones either in synergistic or antagonistic relationships.

Chloroplast development and biogenesis are among the major events in photosynthesis due to their role in light harvesting and subsequent energy conversion, which are tightly regulated by plant hormones [61]. For example, the interaction between auxin and cytokinin ensures optimal chloroplast density and distribution, where cytokinin regulates chloroplast development and plastid division, while auxin regulates cell expansion [60]. The optimal density and proper distribution of chloroplasts significantly influence the photosynthetic process. Synergistic effects between cytokinin (CK) and brassinosteroids (BRs) have also been observed in chloroplast differentiation, thylakoid stacking, and light-harvesting capacity, which ensure efficient photosynthesis under optimal conditions [62–64]. Another interplay between gibberellin (GA) and auxin is involved in maintaining functional chloroplasts. By regulating chloroplast biosynthesis and distribution, phytohormones regulate photosynthesis process in plant.

Chlorophyll biosynthesis is one of the major determinants of efficient photosynthesis, as it is involved in light absorption, energy conversion, and carbon fixation, which are regulated by synergistically by CK, GA, and BR [65, 66]. For example, CK enhances the expression of genes responsible for chlorophyll biosynthesis, GA promotes chloroplast development, and BR ensures the stability of photosystem-related proteins. The auxin–CK interaction tightly regulates chlorophyll biosynthesis, with auxin fine-tuning the CK effect on chlorophyll synthesis. CK, GA, and BR exhibit antagonistic interactions with abscisic acid (ABA) and suppress the inhibitory effect of ABA on chlorophyll biosynthesis under optimal conditions. Another antagonistic relationship is observed between CK/GA/BR and ethylene (ET), where these hormones suppress ET-induced chlorophyll degradation, thereby maintaining chlorophyll biosynthesis under favorable conditions. In summary, CK is one of the major regulators with chlorophyll biosynthesis, however, interplays among different hormones play crucial roles regulation of chlorophyll biosynthesis as well as photosynthesis rate.

Stomata, anatomical structures composed of two guard cells involved in gas exchange and transpiration, are considered major regulators of photosynthesis [67]. Both stomatal movement (opening and closing) and stomatal density play key roles in regulating photosynthesis across plant species. Plant hormones actively regulate stomatal behavior and development, functioning through synergistic or antagonistic interactions [14]. For example, auxin and cytokinin increase the activity of H^+-ATPase in guard cell membranes and enhance K^+ influx, leading to stomatal opening [33]. Under conditions of higher photosynthetic demand, GA works synergistically with CK and indole-3-acetic acid (IAA) to expand stomatal cells and enhance ion transporter activity, thereby maintaining open stomata and improving photosynthesis. Ethylene, through an antagonistic relationship with ABA and a synergistic relationship with auxin, enhances stomatal activity under optimal conditions [68]. Ethylene also downregulates ABA biosynthesis and mitigates its inhibitory effects on stomatal regulation. In summary, phytohormones regulate stomatal movement, distribution, and stomatal size, thus affecting the gas-exchange and regulating photosynthesis rate.

Another aspect of photosynthetic control by phytohormones involves the regulation of enzymes responsible for different steps of photosynthesis. Plant hormones act in both synergistic and antagonistic manners to modulate the activity of photosynthesis-related enzymes, such as Rubisco [69]. For example, auxin, in synergy with CK and GA, enhances Rubisco activity [7]. The combination of IAA, GA, and CK promotes photosynthetic enzyme activity and improves photosynthetic efficiency. In some cases, GA inhibits ABA activity while enhancing CK activity, thereby promoting photosynthesis under normal conditions [70]. ABA generally exhibits antagonistic interactions with most plant hormones; it suppresses Rubisco biosynthesis and reduces photosynthetic efficiency under non-stress conditions. Jasmonic acid (JA) and salicylic acid (SA) have also been reported to influence photosynthetic enzyme regulation, although their effects are relatively subtle.

4. Molecular and genetic insights of phytohormonal regulation on photosynthesis

Photosynthesis is the central metabolic process that regulates overall plant growth and productivity by providing energy through highly coordinated processes, including chloroplast development, photosynthetic enzyme activity, and gene expression in both the nucleus and chloroplast. Plant hormones regulate and fine-tune photosynthesis at molecular, biochemical, and genetic levels by modulating gene expression, chloroplast biogenesis, stomatal dynamics, and signal transduction cascades.

Chloroplast biosynthesis is controlled by the coordinated expression of nuclear-encoded and plastid-encoded genes, where plant hormones act as key regulators to synchronize these functions. For example, cytokinins (CKs) regulate the synthesis of chlorophyll a/b binding proteins by modulating gene expression and also influence the expression of the gene encoding the Rubisco small subunit [8]. Chlorophyll differentiation is also regulated by CKs. Brassinosteroids (BRs) also promote chloroplast development by modulating transcription factors such as PIF4 and HY5, leading to enhanced photosynthetic activity. By regulating HY5 TF, plant hormones enhance chlorophyll content and thus photosynthesis. Another hormone, indole-3-acetic acid (IAA), influences plastid division and differentiation by regulating gene expression and enhancing the expression of the HY5 transcription factor, ultimately affecting photosynthesis. HY5 is the crucial transcriptional factor regulating the genes contributing to photosynthetic pigment biosynthesis and photosystem repair [71]. CKs are major regulators of chloroplast differentiation and maintenance through the activation of nuclear genes encoding photosynthetic proteins. Two key genes, *IPT* (isopentenyltransferase) and *CKX* (cytokinin oxidase/dehydrogenase), regulate CK biosynthesis and degradation, respectively. Changes in the expression of these genes significantly affect photosynthesis by influencing chlorophyll content, leaf senescence, and Rubisco accumulation [72, 73]. For example, overexpression of *IPT* gene in tomato results in higher accumulation of cytokinin associated with delayed leaf senescence and increased photosynthesis rate [72]. On the other hand, mutation in one of the *CKX* gene in maize results in downregulation of photosynthesis related genes [73]. These studies suggest direct control of CKs in photosynthesis and chlorophyll biosynthesis related gene expression in regulation of the photosynthesis rate.

Phytohormones regulate both the light-harvesting complexes and Calvin cycle enzymes by influencing transcriptional regulation. For instance, under abiotic stress, abscisic acid (ABA) activates *REB/ABF*, *ABI5*, and *SnRK2* kinases, which regulate the expression of stress-protective chloroplast genes to stabilize thylakoid membranes [74]. Another aspect of ABA-induced regulation on photosynthesis is through the modulation of D1 repair cycle, which assists photosystem II recovery after stress conditions. For example, a mutant study in rice flag leaf suggest that ABA is involved in regulation of genes involved in D1 protein repair turnover and PSII repair cycle [75]. BRs also contribute to enhance PSII stability by positive regulation on D1 protein repair. BRs enhance both PSII and PSI efficiency by increasing D1 protein synthesis, which improves electron transport and stabilizes thylakoid membranes [76, 77]. ABA regulated stability in PSII has been documented through the regulation of NCED enzymatic pathway; however, it leads to limited CO_2 assimilation due to stomatal closure. ABA perception occurs via PYR/PYL/RCAR receptors, which inhibit PP2C phosphatases and activate SnRK2 kinases. These kinases phosphorylate ABA-responsive transcription factors that regulate chloroplast-related stress-response genes [78]. Another major pathway of phytohormonal regulation on photosynthesis is through the regulation of DELLA proteins. Some plant hormones such as GAs promote the expression of chlorophyll biosynthetic genes through DELLA protein degradation and inhibit chlorophyll breakdown. Several genes are involved in GA biosynthesis and catabolism, directly influencing photosynthesis [79].

Auxin regulates the photosynthesis process by regulating the genes involved in chloroplast differentiation. The auxin response pathway involves AUX/IAA receptors and ARF transcription factors, which regulate genes responsible for chloroplast differentiation. Genetic evidence shows that arf2 mutants have reduced chlorophyll content and impaired light-harvesting complex formation, whereas YUC overexpression increases chloroplast numbers and chlorophyll biosynthesis. In another study, overexpression of ARF6 in tomato results in improved photosynthesis and sugar accumulation with increased auxin activity, suggesting its role in photosynthesis regulation through the enhanced chlorophyll biosynthesis and chloroplast development. Unlike IAA, ABA, and BRs, other plant hormones such as SA, JA, and ET pathways are primarily involved in redox regulation and photoprotection. *ICS1*, a key

Phytohormone	Responsible gene(s)	Role of hormones on gene expression	References
Auxin	*CAB*, *psaA*, *psbA*, *RbcS*	Improves chloroplast structure and regulates light-harvesting proteins	Salazar-Iribe and De-la-Peña [18]
Cytokinin	*RbcL*, *RbcS*, *CAB*, *petB*, *ndhB*	Significantly elevates photosynthesis genes expression and delays senescence process	Cortleven and Schmülling [8]
Gibberellins	*GASA*, *XTH*, *RCA*	Enhances growth and promotes Calvin cycle related enzyme activity	Yuan and Xu[82]
Abscisic acid	*NCED3*, *RD29B*, *RbohF*, *psbA*	Suppresses PSII related genes expression and resilience stress response	Yoshida et al. [83]
Ethylene	*ACS*, *ERF1*, *SAG12*, *CAB*	Suppresses genes expression associated with photosynthesis during senescence	Xie et al. [84]
Brassinosteroids	*BZR1*, *RbcL*, *LHCB*, *FNR*, *psaB*	Elevates Calvin cycle and electron transport associated genes expression	Lv and Li [85]
Salicylic Acid	*PR1*, *psbA*, *CAB*, *RbcS*, *AOX1a*	Boosts photosynthesis efficiency	Miura and Tada [86]
Jasmonic Acid	*JAZ*, *VSP2*, *MYC2*, *CAB*	Suppresses photosynthetic associated gene expression	Zhang et al. [87]
Melatonin	*RbcL*, *psbA*, *CAB*, *APX*, *SOD*	Upregulates photosynthesis gene expression	Liang et al. [88]

Table 2.
Phytohormones induced genes regulate photosynthesis process in plant.

gene in SA biosynthesis, targets transcription factors that regulate antioxidant and PSII repair genes [80]. Mutation in SA and JA biosynthesis pathways resulted in lower accumulation of these two hormones, associated with increased phototoxic effect on photosystem [81]. Hormone-induced genes controlling photosynthesis are presented in **Table 2**.

5. Phytohormones and photosynthesis under abiotic stress conditions

Being sessile organisms, plants are continuously exposed to a wide range of unfavorable environmental conditions and experience various abiotic stresses such as drought, extreme temperatures (both high and low), salinity, nutrient deficiency, waterlogging, and heavy metal toxicity. These abiotic stresses adversely affect all aspects of plant growth, development, and yield by disrupting physiological and biochemical processes [76]. Among these, photosynthesis is one of the most crucial and vulnerable physiological processes impacted by abiotic stresses. Stress conditions influence almost every event associated with photosynthesis. For example, abiotic stress leads to chlorophyll degradation, inhibition of photosynthetic enzymes, and excessive production of reactive oxygen species (ROS), ultimately reducing photosynthetic efficiency across many crop species [89]. However, plant hormones play crucial roles in maintaining and regulating photosynthesis under abiotic stress conditions.

Among the different phytohormones, abscisic acid (ABA) is a central regulator in controlling both stomatal activity and photosynthesis under stress conditions [90]. Abiotic stresses such as drought and salinity enhance ABA biosynthesis in photosynthetic tissues, leading to stomatal closure and reduced CO_2 uptake. At the same time, ABA upregulates antioxidant enzymes such as superoxide dismutase (SOD), catalase (CAT), and ascorbate peroxidase (APX), which protect photosystem II (PSII) from oxidative damage [91]. Another aspect of ABA-regulated photosynthesis is its role in the regulation of late embryogenesis abundant (LEA) proteins that help stabilize thylakoid membranes under stress [92]. ABA-induced transcriptional regulation also contributes to photosynthetic maintenance through interactions with SnRK2 and AREB/ABF transcription factors, enhancing the expression of photosynthesis-related genes. Exogenous application of ABA further supports its role in regulating stomatal behavior and photosynthesis under abiotic stress [91]. In summary, ABA plays crucial roles in activation of enzymatic activity, ROS detoxification, protection PSII, and regulate stomatal features, resulting in improved photosynthesis under abiotic stress conditions.

One of the most detrimental effects of abiotic stress is the degradation of photosynthetic pigments and the disruption of the photosynthetic machinery, including chloroplast ultrastructure. Under such conditions, cytokinins (CKs) play an important role in preventing pigment degradation and maintaining chloroplast integrity. For example, CKs delay leaf senescence by stabilizing chlorophyll biosynthetic proteins [93]. During drought, heat, or salinity stress, CK levels in leaves typically decline, leading to accelerated senescence. However, elevated CK levels maintain Rubisco activity and preserve chloroplast ultrastructure, thereby sustaining photosynthesis [94]. Exogenous application of CKs has been shown to protect Rubisco and maintain chloroplast stability under stress conditions [95, 96]. CKs also exhibit an antagonistic relationship with ABA, promoting stomatal opening and maintaining a balance between water retention and gas exchange.

Jasmonic acid (JA) levels generally increase under abiotic stresses such as drought and salinity. JA modulates chlorophyll degradation and adjusts photosynthetic activity accordingly [97]. In addition, JA enhances the antioxidant defense system by up-regulating enzymes that mitigate ROS overproduction under stress. For example, exogenous application of JA enhanced the antioxidative activity and improved photosynthesis rate in tomato [98]. Balancing growth and defense is a major challenge during abiotic stress, and JA, through interactions with salicylic acid (SA) and ethylene (ET), helps coordinate this balance to optimize photosynthesis and stress tolerance. Another major hormone, auxin, contributes to chloroplast biogenesis through transcriptional regulation. It also indirectly influences photosynthesis by regulating leaf morphology, vein patterning, and photosynthetic surface area [17]. Photosynthetic acclimation is one of the major adaptive strategies plants employ under stress conditions. Exogenous application of IAA enhances the photosynthetic features under different abiotic stress conditions as summarized by Swain et al. [99]. Auxin, in coordination with CKs and brassinosteroids (BRs), plays a key role in maintaining photosynthetic acclimation during abiotic stress [100]. In addition to ABA, auxin, and cytokinin, ethylene is major responsive plant hormones in regulating photosynthesis under abiotic stress conditions, having two types of roles. Low levels of ET promote photosynthetic enzyme activity and chloroplast development, whereas stress-induced increases in ET often accelerate leaf senescence and chlorophyll degradation [101]. However, in certain stress scenarios, Et also contributes to stress tolerance. For example, enhanced ET levels under waterlogging conditions induce aerenchyma formation and regulate carbohydrate metabolism, helping to sustain photosynthetic activity [102].

In addition to the classical well-known hormones, newly characterized phytohormones, such as melatonin (MT) and strigolactones (SLs), have been shown to regulate photosynthesis under stress conditions [103, 104]. Under abiotic stress conditions, MT protects the photosynthetic pigment, including chlorophyll a, b, and total chlorophyll, and enhances the net photosynthesis rate as observed under salinity stress [105]. Similarly, SLs regulate leaf area, stomatal conductance, and chlorophyll retention under abiotic stresses such as nutrient deficiency and drought [104]. Several studies reveal their exogenous application potential in mitigating the abiotic stress-induced damages in plants by enhancing the antioxidant activity, chlorophyll retention, and improved photosynthesis.

6. Future perspectives

Phytohormones play crucial roles in photosynthesis by influencing chloroplast development, stomatal regulation, enzymatic activity, and gene expression. They act in an integrated and interdependent manner, particularly under stress conditions, making photosynthesis one of the most complex physiological processes that requires further study to expand our current understanding. One key area for future research is the interactive networks and crosstalk among phytohormones, including their synergistic and antagonistic relationships under single and combined abiotic stresses. Secondly, the molecular regulation of photosynthesis-related genes by phytohormones under stress conditions remains poorly understood. Thirdly, the effects of phytohormones on the structural integrity and functional efficiency of photosystems I and II (PSI and PSII) under stress require more evidence-based investigation. Fourthly, hormonal regulation is often dose-dependent, highlighting the need for systematic experimental studies to determine optimal concentrations for enhancing photosynthesis. Finally, the role of phytohormones in mitigating stress-induced reactive oxygen species (ROS) through enzymatic activity and membrane stabilization, thereby improving photosynthetic performance, warrants further research. Understanding these aspects will facilitate the targeted use of phytohormones in stress mitigation strategies, ultimately contributing to sustainable crop production under future climate conditions.

7. Conclusion

In summary, the close interaction between phytohormones and photosynthesis is a vital component of plant performance under both optimal and stress environments. Phytohormones namely CKs, auxin, GAs and so on play crucial roles in enhancing photosynthetic efficiency by modulating stomatal conductance, chlorophyll synthesis, and regulating the chlorophyll and photosynthetic genes expression; thus these activities help to boost the robust growth and development of plants. Conversely, during their life cycle, plants face different kinds of environmental stresses like extreme temperature, water scarcity, salinity and it is assumed that phytohormonal regulation shifts to enhance stress tolerance. Hormone like ABA and ethylene orchestrating the wide range of adaptive responses; ABA induces stomatal closure to preserve water, whereas ethylene triggers stress responsive genes expression that increases cellular protection mechanisms. Moreover, JAs and SA have the potential roles to modify photosynthesis processes under stress conditions, which highlight the complexity and multifaceted nature of this regulatory network. Understanding the nuanced mechanisms by which phytohormones influence photosynthesis under varying environmental conditions offers valuable insights for developing resilient crops. By leveraging this knowledge, it may be possible to engineer plants with enhanced photosynthetic efficiency and stress tolerance, thereby contributing to sustainable agriculture in the face of a changing climate. Future research should continue to explore the crosstalk between different phytohormonal pathways and their integration with other signaling networks, as this will undoubtedly reveal new strategies for optimizing plant productivity and resilience.

Acknowledgements

The authors acknowledge the use of ChatGPT for language editing in the chapter.

Author details

Abu Bakar Siddique[1], Ayesha Siddiqua[2], Mst. Mahzabin Zannat[2] and Mohammad Shah Jahan[2]*

1 Tasmanian Institute of Agriculture, University of Tasmania, Launceston, Tasmania, Australia

2 Key Laboratory of Stress Physiology and Molecular Biology, Department of Horticulture, Sher-e-Bangla Agricultural University, Dhaka, Bangladesh

*Address all correspondence to: shahjahansau@gmail.com

References

[1] Häder DP. Photosynthesis in plants and algae. Anticancer Research. 2022;**42**(10):5035-5041. DOI: 10.21873/anticanres.16012

[2] Zhu XG, Long SP, et al. Improving photosynthetic efficiency for greater yield. Annual Review of Plant Biology. 2010;**61**:235-261

[3] Ort DR, Merchant SS, et al. Redesigning photosynthesis to sustainably meet global food and bioenergy demand. Proceedings of the National Academy of Sciences of the United States of America. 2015;**112**:8529-8536

[4] Simkin AJ, Faralli M, Ramamoorthy S, Lawson T. Photosynthesis in non-foliar tissues: Implications for yield. The Plant Journal. 2020;**101**(4):1001-1015

[5] Leister D. Enhancing the light reactions of photosynthesis: Strategies, controversies, and perspectives. Molecular Plant. 2023;**16**(1):4-22. DOI: 10.1016/j.molp.2022.08.005

[6] Li YT, Gao HY, Zhang ZS. Effects of environmental and non-environmental factors on dynamic photosynthetic carbon assimilation in leaves under changing light. Plants. 2023;**12**(10):2015. DOI: 10.3390/plants12102015

[7] Hudeček M, Nožková V, et al. Plant hormone cytokinin at the crossroads of stress priming and control of photosynthesis. Frontiers in Plant Science. 2023;**13**:1103088. DOI: 10.3389/fpls.2022.1103088

[8] Cortleven A, Schmülling T. Regulation of chloroplast development and function by cytokinin. Journal of Experimental Botany. 2015;**66**(6):4999-5013. DOI: 10.1093/jxb/erv132

[9] Khan NA, Ferrante A, Munné-Bosch S. The Plant Hormone Ethylene: Stress Acclimation and Agricultural Applications. Cambridge (MA): Elsevier; 2022. p. 248

[10] Lima VF, Medeiros DB, et al. Toward multifaceted roles of sucrose in the regulation of stomatal movement. Plant Signaling & Behavior. 2018;**13**(8):e1494468. DOI: 10.1080/15592324.2018.1494468

[11] Siddiqui H, Hayat S, Bajguz A. Regulation of photosynthesis by brassinosteroids in plants. Acta Physiologiae Plantarum. 2018;**40**(59):1-17

[12] He Y, Fukushige H, et al. Evidence supporting a role of jasmonic acid in Arabidopsis leaf senescence. Plant Physiology. 2002;**128**(3):876-884

[13] Sharma A. The role of salicylic acid in plants exposed to heavy metals. Plants. 2020;**9**(11):1479

[14] Wei H, Jing Y, Zhang L, Kong D. Phytohormones and their crosstalk in regulating stomatal development and patterning. Journal of Experimental Botany. 2021;**72**(7):2356-2370. DOI: 10.1093/jxb/erab034

[15] Li L, Cheng X, et al. Exogenous melatonin improves peanut field productivity and quality at reduced nitrogen application. Field Crops Research. 2024;**319**:109650

[16] Gururani MA, Mohanta TK, Bae H. Current understanding of the interplay between phytohormones and photosynthesis under environmental stress. International Journal of Molecular Sciences. 2015;**16**(8):19055-19085. DOI: 10.3390/ijms160819055

[17] Zhang Q, Gong M, et al. Roles of auxin in the growth, development, and stress tolerance of horticultural plants. Cells. 2022;**11**(17):2761. DOI: 10.3390/cells11172761

[18] Salazar-Iribe A, De-la-Peña C. Auxins, the hidden player in chloroplast development. Plant Cell Reports. 2020;**39**:1595-1608. DOI: 10.1007/s00299-020-02596-y

[19] Xiong Y, Jiao Y. The diverse roles of auxin in regulating leaf development. Plants. 2019;**8**(7):243

[20] Gao J, Zhuang S, Zhang W. Advances in plant auxin biology: Synthesis, metabolism, signaling, interaction with other hormones, and roles under abiotic stress. Plants (Basel). 2024;**13**(17):2523. DOI: 10.3390/plants13172523

[21] Zhao Y. Auxin biosynthesis and its role in plant development. Annual Review of Plant Biology. 2010;**61**:49-64

[22] Tivendale ND, Millar AH. How is auxin linked with cellular energy pathways to promote growth? New Phytologist. 2022;**233**(6):2397-2404. DOI: 10.1111/nph.17946

[23] Jiang Y, Zhang J. Water stress-induced abscisic acid accumulation triggers the generation of reactive oxygen species in maize leaves. Journal of Experimental Botany. 2002;**53**(379):2401-2410

[24] Acharya BR, Assmann SM. Hormonal regulation of stomatal function. Plant Molecular Biology. 2009;**69**(4):451-462

[25] Brenner WG, Schmülling T. Transcript profiling of cytokinin action in Arabidopsis roots and shoots discovers largely similar but also organ-specific responses. BMC Plant Biology. 2012;**12**:112

[26] Vylíčilová H, Husičková A, Spíchal L, Srovnal J, Doležal K, Plíhal O, et al. C2-substituted aromatic cytokinin sugar conjugates delay the onset of senescence by maintaining the activity of the photosynthetic apparatus. Phytochemistry. 2016;**122**:22-33

[27] Muniandi SK, Hossain MA, et al. Gibberellic acid (GA3) affects growth and development of some selected kenaf (*Hibiscus cannabinus* L.) cultivars. Industrial Crops and Products. 2018;**118**:180-187. DOI: 10.1016/j.indcrop.2018.03.036

[28] Xie J, Tian J, et al. Association genetics and transcriptome analysis reveal a gibberellin-responsive pathway involved in regulating photosynthesis. Journal of Experimental Botany. 2016;**67**(11):3325-3338

[29] Afroz S, Mohammad F, et al. Exogenous application of gibberellic acid counteracts the ill effect of sodium chloride in mustard. Turkish Journal of Biology. 2005;**29**(4)

[30] Sakamoto T, Miura K, et al. An overview of gibberellin metabolism enzyme genes and their related mutants in rice. Plant Physiology. 2004;**134**(4):1642-1653. DOI: 10.1104/pp.103.033696

[31] Raza A, Charagh S, et al. Jasmonic acid: A key frontier in conferring abiotic stress tolerance in plants. Plant Cell Reports. 2021;**40**(8):1513-1541. DOI: 10.1007/s00299-020-02614-z

[32] Gao S, Gao J, et al. ABF2, ABF3, and ABF4 promote ABA-mediated chlorophyll degradation and leaf senescence by transcriptional activation of chlorophyll catabolic genes and senescence-associated genes in Arabidopsis. Molecular Plant. 2016;**9**(9):1272-1285. DOI: 10.1016/j.molp.2016.06.006

[33] Vate'n A, Soyars CL, et al. Modulation of asymmetric division diversity through cytokinin and speechless regulatory interactions in the Arabidopsis stomatal lineage. Developmental Cell. 2018;**47**:53-66. DOI: 10.1016/j.devcel.2018.08.007

[34] Chen L, Dodd IC, et al. Ethylene limits abscisic acid- or soil drying-induced stomatal closure in aged wheat leaves. Plant Cell Environment. 2013;**36**:1850-1859

[35] Hayat S, Ahmad A, Ahmad A. Brassinosteroids: Bioactivity and Crop Productivity. Dordrecht, The Netherlands: Springer Science+Business Media Dordrecht; 2003

[36] Xia XJ, Wang YJ, et al. Reactive oxygen species are involved in brassinosteroid-induced stress tolerance in cucumber. Plant Physiology. 2009;**150**(2):801-814

[37] Yu J, Hunag LF, et al. A role for brassinosteroids in the regulation of photosynthesis in cucumber. Journal of Experimental Botany. 2004;**55**(399):1135-1143

[38] Hayat S, Mori M, et al. Physiological role of brassinosteroids: An update. Indian Journal of Plant Physiology. 2010;**15**(2)

[39] Guo F, Guo J, Shen Y. Advances from chlorophyll biosynthesis to photosynthetic adaptation, evolution and signaling. Plant Stress. 2024;**12**:100470. DOI: 10.1016/j.stress.2024.100470

[40] Li T, Hu Y, et al. Salicylic acid alleviates the adverse effects of salt stress in *Torreya grandis* cv. Merrillii seedlings by activating photosynthesis and enhancing antioxidant systems. PLoS One. 2013;**8**(11):e84601

[41] Kang G, Li G, et al. Proteomics reveals the effects of salicylic acid on growth and tolerance to subsequent drought stress in wheat. Journal of Proteome Research. 2014;**13**(12):5175-5190

[42] Nazar R, Umar S, et al. Salicylic acid supplementation improves photosynthesis and growth in mustard (*Brassica juncea* L.) through modulation of proline metabolism under salt stress. Plant Physiology and Biochemistry. 2015;**94**:217-226

[43] Ali S, Baek KH. Jasmonic acid signaling pathway in response to abiotic stresses in plants. International Journal of Molecular Sciences. 2020;**21**(2):621

[44] Ananiev ED, Ananieva K, Todorov I. Effect of methyl ester of jasmonic acid, abscisic acid and benzyladenine on chlorophyll synthesis in excised cotyledons of *Cucurbita pepo* (zucchini). Bulgarian Journal of Plant Physiology. 2004;**30**:51-63

[45] Chen J, Yan Z, et al. Effect of methyl jasmonate on cadmium uptake and antioxidativeia obovata seedlings under cadmium stress. Ecotoxicology and Environmental Safety. 2014;**104**:349-356

[46] Yan Z, Zhang W, et al. Methyl jasmonate alleviates cadmium toxicity in *Solanum nigrum* by regulating metal uptake and antioxidative capacity. Biologia Plantarum. 2015;**59**:373-381

[47] Kovac M, Ravnikar M. The effect of jasmonic acid on the photosynthetic pigments of potato plant grown in vitro. Plant Science. 1994;**103**:11-17

[48] Asma M, Lingakumar K. Jasmonate foliar spray induced vegetative growth and pigment composition in mungbean (*Vigna Radiata L.*) WILCZEK. International Journal of Advanced Research. 2015;**3**(1):664-669

[49] Arnao MB, Hernández-Ruiz J. Protective effect of melatonin against chlorophyll degradation during the senescence of barley leaves. Journal of Pineal Research. 2009;**46**(1):58-63

[50] Wang M, Zhang S, Ding F. Melatonin mitigates chilling-induced oxidative stress and photosynthesis inhibition in tomato plants. Antioxidants (Basel). 2020;**9**:218

[51] Khan NH, Nafees M, Zuljalal F, Saeed T. Auxin's role in plant development in response to stress. In: Engineering Tolerance in Crop Plants against Abiotic Stress. Boca Raton, FL, USA: CRC Press; 2021. pp. 209-224

[52] Nagar S, Meena S, Karwa S, Dhakar R, Raju D, Kumar S, et al. Regulatory role of cytokinin in abiotic stress tolerance of plants. In: Phytohormones in Abiotic Stress. Boca Raton, FL, USA: CRC Press; 2024. pp. 64-72

[53] Razzaq K, Du J. Phytohormonal regulation of plant development in response to fluctuating light conditions. Journal of Plant Growth Regulation. 2025;**44**(5):1903-1936

[54] Bagal D, Vishnoi R, Rahman A, Sharma V. Role of abscisic acid in abiotic stresses. In: Phytohormones in Abiotic Stress. Boca Raton, FL, USA: CRC Press; 2024. pp. 45-53

[55] Mohorović P, Geldhof B, Holsteens K, Rinia M, Daems S, Reijnders T, et al. Ethylene inhibits photosynthesis via temporally distinct responses in tomato plants. Plant Physiology. 2024;**195**(1):762-784

[56] Wasternack C. A plant's balance of growth and defense–revisited. New Phytologist. 2017;**215**(4):1291-1294

[57] Cheaib A, Killiny N. Photosynthesis responses to the infection with plant pathogens. Molecular Plant-Microbe Interactions. 2025;**38**(1):9-29

[58] Nolan TM, Ukašinović N, et al. Brassinosteroid gene regulatory networks at cellular resolution in the Arabidopsis root. Science. 2023;**379**(6639):eadf4721. DOI: 10.1126/science.adf4721

[59] Hussain A, Faizan M, Ahmed SM. Regulation of photosynthesis by melatonin under optimal and suboptimal conditions. In: Melatonin in Plants: Role in Plant Growth, Development, and Stress Response. Springer Nature Singapore: Singapore; 2024. pp. 35-51

[60] Müller M, Munné-Bosch S. Hormonal impact on photosynthesis and photoprotection in plants. Plant Physiology. 2021;**185**(4):1500-1522. DOI: 10.1093/plphys/kiaa119

[61] Cackett L, Luginbuehl LH, et al. Chloroplast development in green plant tissues: The interplay between light, hormone, and transcriptional regulation. New Phytologist. 2021;**233**(5):2000-2016. DOI: 10.1111/nph.17839

[62] Cortleven A, Marg I, et al. Cytokinin regulates the etioplast-chloroplast transition through the two-component signaling system and activation of chloroplast-related genes. Plant Physiology. 2016;**172**(1):464-478. DOI: 10.1104/pp.16.00640

[63] Okazaki K, Kabeya Y, et al. The Plastid division 1 and 2 components of the chloroplast division machinery determine the rate of chloroplast division in land plant cell differentiation. Plant Cell. 2009;**21**:1769-1780

[64] Cortleven A, Schmülling T. Regulation of chloroplast development and function by cytokinin. Journal of

Experimental Botany. 2015;**66**(16):4999-5013. DOI: 10.1093/jxb/erv132

[65] Liu X, Li Y, Zhong S. Interplay between light and plant hormones in the control of Arabidopsis seedling chlorophyll biosynthesis. Frontiers in Plant Science. 2017;**8**:1433. DOI: 10.3389/fpls.2017.01433

[66] Wang X, Shi J. Leaf chlorophyll content is the crucial factor for the temporal and spatial variation of global plants leaf maximum carboxylation rate. Science of The Total Environment. 2024;**927**:172280. DOI: 10.1016/j.scitotenv.2024.172280

[67] Ma Y, Yuan Z, et al. Stomatal and non-stomatal regulations of photosynthesis in response to salinity, and K and Ca fertigation in cotton (*Gossypium hirsutum* L cv.). Environmental & Experimwntal Botany. 2025;**230**:106092. DOI: 10.1016/j.envexpbot.2025.106092

[68] Tanaka Y, Sano T, et al. Ethylene inhibits abscisic acid-induced stomatal closure in Arabidopsis. Plant Physiology. 2005;**138**(4):2337-2343. DOI: 10.1104/pp.105.063503

[69] Amaral J, Lobo AKM, Carmo-Silva E. Regulation of rubisco activity in crops. The New Phytologist. 2024;**241**(1):35-51. DOI: 10.1111/nph.19369

[70] Morris B, Khan S, et al. Gibberellic acid reverses glucose-inhibited photosynthesis and growth via reduced glutathione and hormonal crosstalk in chromium-treated mustard. Plant Stress. 2024;**14**:100694. DOI: 10.1016/j.stress.2024.100694

[71] Gangappa SN, Botto JF. The multifaceted roles of HY5 in plant growth and development. Molecular Plant. 2016;**9**(10):1353-1365. DOI: 10.1016/j.molp.2016.07.002

[72] Glanz-Idan N, Lach M, Tarkowski P, Vrobel O, Wolf S. Delayed leaf senescence by upregulation of cytokinin biosynthesis specifically in tomato roots. Frontiers in Plant Science. 2022;**13**:922106. DOI: 10.3389/fpls.2022.922106

[73] Wang H, Yan Y, et al. Natural variation in the cytokinin oxidase gene ZmCKX6 influences leaf morphology and yield-related traits in maize. The Plant Journal. 2025;**123**(1):e70332. DOI: 10.1111/tpj.70332

[74] Yamburenko MV, Zubo YO, et al. Abscisic acid represses the transcription of chloroplast genes. Journal of Experimental Botany. 2013;**64**(4):4491-4502. DOI: 10.1093/jxb/ert258

[75] Wang F, Liu J, et al. Involvement of abscisic acid in psii photodamage and d1 protein turnover for light-induced premature senescence of rice flag leaves. PLoS One. 2016;**11**(8):e0161203. DOI: 10.1371/journal.pone.0161203

[76] Chauhan J, Prathibha MD, et al. Plant photosynthesis under abiotic stresses: Damages, adaptive, and signaling mechanisms. Plant Stress. 2023;**10**:100296. DOI: 10.1016/j.stress.2023.100296

[77] Hu X, Qian Q, et al. The U-box E3 ubiquitin ligase TUD1 functions with a heterotrimeric g α subunit to regulate brassinosteroid-mediated growth in rice. PLoS Genetics. 2013;**9**(3):e1003391. DOI: 10.1371/journal.pgen.1003391

[78] Huang Y, Guo Y, et al. 9-cis-epoxycarotenoid dioxygenase 3 regulates plant growth and enhances multi-abiotic stress tolerance in rice. Frontiers in Plant Science. 2018;**9**:162. DOI: 10.3389/fpls.2018.00162

[79] Islam S, Park K, et al. Structural insights into gibberellin-mediated

DELLA protein degradation. Molocular Plant. 2025;**18**(7):1210-1221. DOI: 10.1016/j.molp.2025.06.010

[80] Duan J, Lee KP, et al. Impaired PSII proteostasis promotes retrograde signaling via salicylic acid. Plant Physiology. 2019;**180**(4):2182-2197. DOI: 10.1104/pp.19.00483

[81] Browse J. The power of mutants for investigating jasmonate biosynthesis and signaling. Phytochemistry. 2009;**70**(13-14):1539-1546. DOI: 10.1016/j.phytochem.2009.08.004

[82] Yuan L, Xu DQ. Stimulation effect of gibberellic acid short-term treatment on leaf photosynthesis related to the increase in Rubisco content in broad bean and soybean. Photosynthesis Research. 2001;**68**:39-47. DOI: 10.1023/A:1011894912421

[83] Yoshida T, Mogami J, Yamaguchi-Shinozaki K. ABA-dependent and ABA-independent signaling in response to osmotic stress in plants. Current Opinion in Plant Biology. 2014:133-139. DOI: 10.1016/j.pbi.2014.07.009

[84] Xie XL, Xia XJ, et al. A novel ethylene responsive factor CitERF13 plays a role in photosynthesis regulation. Plant Science. 2017;**256**:112-119. DOI: 10.1016/j.plantsci.2016.11.001

[85] Lv M, Li J. Molecular mechanisms of brassinosteroid-mediated responses to changing environments in Arabidopsis. International Journal of Molecular Sciences. 2020;**21**(8):2737. DOI: 10.3390/ijms21082737

[86] Miura K, Tada Y. Regulation of water, salinity, and cold stress responses by salicylic acid. Frontiers in Plant Science. 2014;**5**:4. DOI: 10.3389/fpls.2014.00004

[87] Zhang ZC, He B, et al. The phytotoxin COR induces transcriptional reprogramming of photosynthetic, hormonal and defence networks in tomato. Plant Biology. 2021;**23**:69-79. DOI: 10.1111/plb.13239

[88] Liang C, Zheng G, et al. Melatonin delays leaf senescence and enhances salt stress tolerance in rice. Journal of Pineal Research. 2015;**59**(1):91-101. DOI: 10.1111/jpi.12243

[89] Muhammad I, Shalmani A, et al. Mechanisms regulating the dynamics of photosynthesis under abiotic stresses. Frontiers in Plant Science. 2021;**11**:615942. DOI: 10.3389/fpls.2020.615942

[90] Lim CW, Baek W, et al. Function of ABA in stomatal defense against biotic and drought stresses. International Journal of Molecular Sciences. 2015;**16**(7):15251-15270. DOI: 10.3390/ijms160715251

[91] Aslam MM, Waseem M, et al. Mechanisms of abscisic acid-mediated drought stress responses in plants. International Journal of Molecular Sciences. 2022;**23**(3):1084. DOI: 10.3390/ijms23031084

[92] Dalal M, Tayal D, et al. Abiotic stress and ABA-inducible Group 4 LEA from Brassica napus plays a key role in salt and drought tolerance. Journal of Biotechnology. 2009;**139**(2):137-145. DOI: 10.1016/j.jbiotec.2008.09.014

[93] Mandal S, Ghorai M, et al. Cytokinin and abiotic stress tolerance - What has been accomplished and the way forward? Frontiers in Genetics. 2022;**13**:943025. DOI: 10.3389/fgene.2022.943025

[94] Efroni I, Han SK, et al. Regulation of leaf maturation by chromatin-mediated

modulation of cytokinin responses. Developmental Cell. 2013;**24**(4):438-445. DOI: 10.1016/j.devcel.2013.01.019

[95] Siddique AB, Shabala S, et al. Reducing heat stress damage in cereal crops through agronomic management and breeding strategies. Plant Stress. 2025;**16**:100888. DOI: 10.1016/j.stress.2025.100888

[96] Zhang Y, Li J, et al. Cytokinin modulates the inhibitory effect of shade stress on photosynthesis, antioxidant capacity and hormone homeostasis to regulate the grain yield in wheat. Frontiers in Plant Science. 2024;**15**:1498123. DOI: 10.3389/fpls.2024.1498123

[97] Rehman M, Saeed MS, Fan X, Salam A, Munir R, Yasin MU, et al. The multifaceted role of jasmonic acid in plant stress mitigation: An overview. Plants. 2023;**12**(23):3982. DOI: 10.3390/plants12233982

[98] Gul N, Masoodi KZ, et al. Study on the impact of exogenously applied methyl jasmonate concentrations on *Solanum lycopersicum* under low temperature stress. BMC Plant Biology. 2023;**23**:437. DOI: 10.1186/s12870-023-04449-8

[99] Swain R, Sahoo S, et al. Instigating prevalent abiotic stress resilience in crop by exogenous application of phytohormones and nutrient. Frontiers in Plant Science. 2023;**14**:1104874. DOI: 10.3389/fpls.2023.1104874

[100] Tian H, Lv B, et al. Auxin-BR interaction regulates plant growth and development. Frontiers in Plant Science. 2018;**8**:2256. DOI: 10.3389/fpls.2017.02256

[101] Chen H, Bullock DA Jr, et al. To fight or to grow: The balancing role of ethylene in plant abiotic stress responses. Plants (Basel). 2021;**11**(1):33. DOI: 10.3390/plants1101003

[102] Husain T, Fatima A, et al. A brief appraisal of ethylene signaling under abiotic stress in plants. Plant Signaling & Behavior. 2020;**15**(9):1782051. DOI: 10.1080/15592324.2020.1782051

[103] Ahmad I, Song X, et al. The role of melatonin in plant growth and metabolism, and its interplay with nitric oxide and auxin in plants under different types of abiotic stress. Frontiers in Plant Science. 2023;**14**:1108507. DOI: 10.3389/fpls.2023.1108507

[104] Khalid MF, Shafqat W, et al. Unveiling the resilience mechanism: Strigolactones as master regulators of plant responses to abiotic stresses. Plant Stress. 2024;**12**:100490. DOI: 10.1016/j.stress.2024.100490

[105] Yin J, Li Y, Han H, Ma J, Liu G, Wu X, et al. Administration of exogenous melatonin improves the diurnal rhythms of the gut microbiota in mice fed a high-fat diet. mSystems. 2020;**5**(3):e00002-e00020. DOI: 10.1128/mSystems.00002-20

Chapter 4

Salt Stress in Crops: Tolerance Mechanisms and Emerging Amelioration Strategies

Kennedy Samuel Banja and Olusola Jacob Oyetunji

Abstract

Salt stress is a deleterious abiotic stress that is silently degrading arable lands, threatening crop production, and compromising food security. Arid and semi-arid regions of the world are the most severely affected and may suffer more as climate change intensifies. Salinity disrupts several levels of plant development, including morphological, physiological disruption, biochemical and molecular effects, resulting in impaired photosynthesis, ion toxicity, and oxidative damage. Several mechanisms for plant survival have been highlighted, including hormonal regulation, antioxidant defense activation, ion homeostasis, and osmotic regulation. Plants adapt to salt stress through the accumulation of compatible solutes, hormonal modulation, antioxidant defense, ion homeostasis, and osmotic adjustment. These adaptation responses vary across species and developmental stages, with early stages of growth being most vulnerable. The development of techniques to mitigate salt stress has expanded beyond traditional breeding to include physiological treatments, such as foliar applications of growth regulators, calcium, nitric oxide, and amino acids, which improves ion balance and oxidative stress tolerance. Biological strategies involving beneficial microbes, such as plant growth-promoting bacteria and arbuscular mycorrhizal fungi, increase nutrient uptake and activate defense mechanisms. Cutting-edge tools such as CRISPR/Cas9 gene editing, nanomaterials, and microbiome engineering are enabling precise, sustainable enhancement of salt tolerance in crops. Although scientific progress has been made, these strategies are inapplicable at the field level, highlighting the need for in-depth, multi-location studies and holistic approaches. The development of resilient cropping systems in saline-prone regions requires a combination of molecular advances with sustainable agronomic practices.

Keywords: salt stress, plant tolerance mechanisms, ion homeostasis, soil amelioration, osmotic regulation, antioxidant defense, hormonal regulation, genetic engineering

1. Introduction

Soil salinity is one of the most serious threats to agriculture in the twenty-first century, silently degrading productive land and compromising crop yields in many parts of the world. At its core, salinity involves the excessive accumulation of soluble salts, particularly sodium (Na^+), chloride (Cl^-), and sulfate (SO_4^{2-}), in the soil. This condition occurs naturally in some arid and semi-arid regions where high evaporation rates leave salts behind as water evaporates. Anthropogenic activities such as excess fertilization and inadequate drainage have further worsened the situation [1–3]. These practices potentially limit the optimum plant development by altering soil chemical balance.

Globally, more than one billion hectares are impacted by soil salinity, posing a considerable threat to food security and sustainable development [1, 3, 4]. In arid regions, climate change is expected to worsen salinization due to heightened droughts and evaporation rates [3–5]. The monitoring and management of soil salinity require remote sensing, soil mapping, and the implementation of sustainable agricultural practices, including enhanced irrigation, drainage, and the cultivation of salt-tolerant crops or beneficial soil microbes [1–3, 6]. Tackling soil salinity is essential for preserving soil fertility, fostering biodiversity, and achieving global sustainability objectives [1, 3, 7].

Despite the far-reaching effects of salinity on plants, the situation is getting serious attention. Breeding salt-tolerant crops and utilizing plant growth-promoting bacteria (PGPB) and fungi are key strategies to reduce salinity and improve crop performance. Satellite imaging and soil salinity mapping have also enhanced the ability to monitor affected regions. Other strategies, including optimized irrigation scheduling, controlled drainage, and the use of organic amendments, are yielding results in ameliorating salinity [1, 6].

2. Morphological effects of salinity on plants

Morphological adaptations are among the earliest and most visible responses of plants to salt stress, and changes in plant structure at any level directed toward conserving water, reducing the accumulation of toxic ions, or maintaining cell integrity. There is a tendency for the root: shoot ratio to increase since it produces deeper roots that can penetrate less saline soil layers, where water is more easily available with reduced salt uptake *via* aerial parts [8, 9]. Root systems become more branched or thicker so that surface area development for selective ion uptake ensures the minimum buildup of toxicity.

Surface area reduction, thickening, and succulence are anatomical variations observed in leaves. Therefore, such features reduce transpiration and help dilute intracellular salts [10, 11]. Some halophytes develop salt glands or bladders that actively excrete extra salts so that cytotoxicity does not take place [12]. In addition, the formation of aerenchyma in the root enables oxygen diffusion under conditions of salinity and hypoxia [11]. More chlorophyll content and waxy cuticles protect the photosynthetic apparatus and help conserve water. The features improve tolerance and are the basic targets for selection when breeding salt-resilient genotypes.

2.1 Growth stage-specific sensitivity to salt stress

Salt affects plants at all stages of growth, but its effects are different across the developmental stage and species. High salt during seed germination and initial seedling growth reduces the rate of germination vigorously as well as the vigor of the seedling and root and shoot development. Therefore, these stages are most sensitive to salinity stress [13–16]. In the vegetative stage, salinity slows down the growth of roots and shoots, shrinks the area of leaves, and disrupts photosynthesis. This is mostly because of osmotic stress and ion toxicity [13, 16].

Salt stress at the reproductive stage may reduce the number of flowers and fruits, pollen viability, and seed set, hence lowering yield in many crops, though this stage is generally less sensitive compared to the vegetative stage [13, 16, 17]. However, salinization of the medium enhances reproductive output in some halophytes, increasing flower number and maintaining or even improving pollen vitality and seed yield due to proper hormonal regulation plus osmolyte accumulation [12, 18]. Salinity can promote sexual reproduction over vegetative growth in some clonal plants, though when exceeding a certain level, it greatly reduces seed production [19]. Tomatoes, for example, are susceptible to salinity stress during the fruiting and flowering stages, which reduce yields by reducing fruit production. However, salt stress may improve fruit quality during this phase [17].

2.2 Physiological effects of salinity on plants

Salinity essentially interferes with the physiology of the plant by perturbing water uptake, nutrient balance, and several related metabolic processes. High salt concentration raises osmotic stress outside the soil surrounding the root zone, making it difficult for the plant to take up water; hence, growth reduction is realized, accompanied by reduced leaf water potential and loss of turgor [20–22]. The toxicity effect brought about by accumulation of sodium (Na^+) and chloride (Cl^-) ions in plant tissues mainly manifests itself through the harmful impacts on enzymatic activities as well as destruction of cellular membranes, besides interference in nutrient uptake, particularly potassium and calcium [9, 16, 22, 23]. Salinity-induced oxidative stress increases the production of reactive oxygen species (ROS) that are damaging to proteins, lipids, and DNA if not immediately neutralized by antioxidants [22, 24]. Energy and growth reduction are mainly imposed by damages related to stomatal closure, chlorophyll degradation, and direct damage imposed on both the structure and function of chloroplasts, wherein these photosynthesis-imposing damages result in inhibition of the process itself, thus leading to reduced energy production [20–22, 25]. The immediate responses include ion compartmentalization, synthesis of osmoprotectants (e.g., proline and glycine betaine), as well as up-regulation of stress-responsive genes and hormones that try to maintain homeostasis within cells while minimizing cellular damage [9, 16, 26]. Long-term and unmitigated salt stress could eventually lead to yield decrease or death, due to leaf senescence, chlorosis, and necrosis [16, 20, 21].

2.3 Biochemical responses to salinity

Salinity stress triggers several biochemical changes in the plant that ultimately lead to the general disruption of cellular activities and threaten growth and productivity. High concentrations of salt will lead to the accumulation of toxic ions (mainly Na^+ and Cl^-) whose presence will disturb ion homeostasis as well as nutrient uptake, and also impose osmotic stress, hence water imbalance [16, 22, 23].

A major biochemical effect is an overproduction of reactive oxygen species (ROS) that can attack proteins, lipids, and nucleic acids, leading to oxidative stress [1, 22]. The plant in return mounts an antioxidant defense in the form of superoxide dismutase, catalase, and peroxidases as well as enzymatic and non-enzymatic antioxidants to quench ROS and protect cellular structures [1, 22, 26, 27]. Also, salinity leads to the accumulation of proline, glycine betaine, and sugars. These compounds assist the stabilization of proteins and membranes, maintenance of osmotic balance, or protection from desiccation [1, 16, 22, 28]. In addition to this, salinity changes the biosynthesis as well as signaling pathways of phytohormones that are controlling stress responses and adaptation mechanisms [1, 22]. Such biochemical changes are often coupled with variations at both gene and protein expression levels that would support plant attempts directed toward toxic ion compartmentalization, metabolic function sustenance, and survival under saline conditions [1, 16, 22, 26].

3. Mechanisms of salt tolerance in plants

Plants set up diverse mechanisms of tolerance to survive and grow well under salt stress, which is mainly an osmotic-plus-ionic-related stress. Upon encountering salt stress, plants face an immediate osmotic stress that inhibits the uptake of water, followed by the gradual accumulation of toxic ions (mainly sodium (Na^+) and chloride ions (Cl^-)). They respond by eliciting integrated morphological, physiological, biochemical, and molecular responses at different levels of organization to minimize injury and maintain metabolism [9, 16, 29].

Regulation of ion homeostasis and transport is the major mechanism for tolerating salt stress in plants. Plants manage salt toxicity by exclusion at the root surface, sequestration of toxic ions into vacuoles through Na^+/H^+ antiporters on tonoplast membranes, and removal from the cytosol by Salt Overly Sensitive 1 (SOS1) channels located on the plasma membrane. At the same time, high-affinity potassium transporters (HKT) play an important role in xylem Na^+ retrieval to avoid its over-accumulation in shoot tissues. H^+-ATPases and H^+-PPases drive these transport processes, and they are controlled by hormones, transcription factors, and post-translational modifications [30–32]. Simultaneously, plants regulate their water balance and maintain cell turgor when faced with osmotic stress through osmotic adjustment, which are accomplished by the accumulation of compatible solutes such as proline, glycine betaine, and soluble sugars. These osmolytes effectively reduce cellular osmotic potential without disrupting normal metabolic processes, stabilize membranes and proteins, and may also function as antioxidants [33–35]. Cotton and barley, for example, use inorganic ions to regulate osmotic conditions, though they must be compartmentalized within the cell to avoid toxicity [36, 37].

The antioxidant system is another major defense mechanism that comes into play against oxidative stress because salinity mainly induces the overproduction of reactive oxygen species (ROS). The enzymatic antioxidants include superoxide dismutase (SOD), catalase (CAT), ascorbate peroxidase (APX), and the non-enzymatic antioxidants, ascorbic acid, glutathione, and phenolic compounds, among others. They neutralize ROS, thus protecting the cellular macromolecules from damage [38, 39]. Their responses are also regulated by phytohormones and genes, in addition to being enhanced by external treatments and by beneficial microbial interactions.

At the molecular level, stress signaling pathways play a great role in coordinating plant responses to salinity. Na^+ sensing and exclusion depend on the SOS pathway, while MAPK cascades and SnRK2 kinases assist signal transduction and activate stress-related genes [18]. These pathways regulate a diverse range of transcription factors, including MYB, WRKY, NAC, AP2/ERF, and bZIP families, which subsequently modulate the expression of genes involved in ion transport, ROS detoxification, osmolyte biosynthesis, and hormonal responses [31, 40, 41]. A balanced stress response is enabled by crosstalk between abscisic acid (ABA), ethylene, jasmonic acid, and salicylic acid [42, 43]. In certain species, especially halophytes, these tolerance traits are expressed more prominently. For instance, salt-tolerant Limonium and Plantago species exhibit improved Na^+ compartmentalization, heightened K+ retention, and increased antioxidant levels [44, 45]. These natural mechanisms provide significant genetic and physiological frameworks for the breeding or engineering of salt-resilient crops.

3.1 Recent advances in salt stress amelioration

In recent years, great progress has been made in the strategies that attenuate the negative impacts of salinity on plant health and productivity. They include physiological treatments and microbial methods, biotechnological innovations, and ecological strategies aimed at increasing plant resistance in salt-affected soils.

A widely adopted approach comprises the foliar application of stress-mitigating compounds that include the plant growth regulators, amino acids (mainly proline and aspartic acid), and signaling molecules such as calcium, nitric oxide, and salicylic acid. Studies have indicated that photosynthetic performance, ion homeostasis or balance, and antioxidant defense of plants under saline conditions can be improved by the application of these compounds [46–49]. A foliar treatment of indole-3-acetic acid (IAA) on faba bean and gibberellin on papaya reduced growth inhibition as well as oxidative damage. Similarly, exogenous proline and aspartic acid application have raised osmolytes and antioxidant activities in cereals like wheat and maize [48, 50].

Calcium represents supplementation as an effective approach. It plays a protective role by mitigating ionic toxicity as well as oxidative damage. Calcium has proven results in improving photosynthetic efficiency, nutrient uptake, and fruit quality under saline conditions for tomatoes [49]. At the same time, nitric oxide has recently been considered a very effective signaling molecule in increasing antioxidant capacity while reducing the accumulation of Na^+ and increasing photosynthetic efficiency and nutrient uptake in crops such as wheat and lettuce [46, 51]. Similarly, Chinese cabbage had a higher phenolic compound accumulation and improved photosynthesis under salinity stress when treated with salicylic acid and ferulic acid [52].

The use of plant growth-promoting bacteria (PGB) and arbuscular mycorrhizal fungi (AMF) in salt stress amelioration is increasingly considered [53]. These organisms improve plant salt tolerance through the modulation of antioxidant enzymes, nutrient uptake improvement, as well as expression regulation of stress-responsive genes. The application of *Bacillus thuringiensis* PM25 to maize plants resulted in decreased oxidative damage while simultaneously modulating growth-related genes and enhancing leaf pigment production under the salt stress conditions [54]. *Trichoderma asperellum* enhances salt tolerance in tomato seedlings through the activation of antioxidant systems and stress-resistance gene upregulation according to Hu et al. [55]. These microbial agents generate hormones and metabolites that improve root structures and boost osmolyte levels, making them effective, sustainable bio-tools for salinity management [56, 57].

An additional developing field for salt stress mitigation is nanotechnology. Scientists have used proline-functionalized carbon quantum dots to treat grapevine plants, resulting in a higher antioxidant activity along with better photosynthetic performance and increased stress resistance by the plants [58]. Natural extracts derived from the brown macroalga *Sargassum wightii* function as a critical source of bioactive compounds, which help restore metabolic balance while maintaining ionic stability during salt stress conditions [59].

Biotechnological advancements such as genetic engineering and CRISPR/Cas9 genome editing have expanded genetic resources for crop salt tolerance enhancement into complex new dimensions. Scientific efforts initially focused on enhancing the expression of ion transport genes (NHX, SOS1) alongside those involved in antioxidant defenses and osmoprotectant production. Current scientific methods employ CRISPR/Cas9 to remove the negative regulators or precisely alter key genes while maintaining yield stability. The experimental deletion of HyPRP1 in tomato, together with GmAITR genes in soybean, resulted in enhanced salt tolerance across both laboratory tests and field conditions [60, 61].

Rice and wheat show similar advancements where manipulation of transcription factors alongside hormone-related genes has boosted stress resistance [62, 63].

The molecular architecture underlying salt tolerance has been elucidated through the advancements in genomics, proteomics, and metabolomics, which operate alongside these initiatives. Through these methodologies, researchers have discovered novel transcription factors alongside gene families and metabolic pathways that govern ion transport as well as ROS detoxification and stress adaptation processes [64–67]. The current use of data-rich platforms enables the development of stress-resilient crops through precise breeding and engineering techniques.

The concept of microbiome engineering, which involves altering or adding beneficial rhizosphere or endophytic microbes, offers a sustainable and scalable method for salinity management. Through the modification of soil microbial populations, plants can develop enhanced stress resistance while reducing ROS levels and maintaining growth in saline conditions [57, 68].

3.1.1 Prospect and research gaps

Despite major progress in understanding salt stress responses and developing mitigation approaches, numerous essential gaps and obstacles persist, particularly in applying this knowledge to create practical field solutions. The main issue lies in the fact that many desirable traits and interventions discovered under the controlled laboratory or greenhouse conditions fail to deliver reliable results when applied in real-world fields where complex interactions between biotic and abiotic stressors occur [69, 70]. The discrepancy highlights an urgent requirement for field validation of salt-tolerant genotypes and treatment strategies across multiple environments. The intricate polygenic characteristics of salt tolerance remain an unresolved scientific challenge [9, 70]. The intricate balance between growth and defense mechanisms remains poorly understood, which limits the practical use of specific tolerance traits.

The understanding of natural genetic variation for salt tolerance among underutilized crops and wild relatives of major staple crops remains significantly incomplete. These resources remain largely unexploited, despite exhibiting promising traits such as efficient Na^+ exclusion, increased osmolyte accumulation, and resilient root systems [71]. Integrating genetic reservoirs into breeding programs through techniques like introgression and genomic-assisted selection presents novel methods to boost crop resilience.

Despite extensive characterization of major signaling pathways such as SOS, mitogen-activated protein kinase (MAPK), and SnRK2 cascades at the molecular level, fundamental questions remain about how plants initially detect salt stress. The primary sensors' identities and their complex interactions with hormonal and redox pathways remain undefined [69, 72]. Post-translational mechanisms, epigenetic controls, and circadian rhythm interactions represent emerging stress response components that demand additional scientific investigation [73, 74].

Even though scientific recognition of plant-associated microbiomes' beneficial functions is gaining traction, research within this field remains nascent. Scientific understanding remains scarce regarding the molecular interactions through which rhizosphere and endophytic microbes modulate plant salt stress responses [68, 75]. Transforming agricultural stress management through tailored microbial consortia and microbiome engineering platforms requires an intricate systems-level comprehension of plant-microbe-soil interactions under salt stress conditions.

Combined with machine learning and predictive modeling, multi-omics technologies such as transcriptomics, proteomics, metabolomics, and phenomics will be crucial. In addition to deciphering complex genotype-phenotype-environment relationships, these tools can help identify new regulatory hubs and prioritize breeding targets [65, 70]. In the end, these insights should be matched with participatory breeding programs and inclusionary policies that promote salt-resilient varieties within vulnerable farming communities.

4. Conclusion

Crop production is still constrained by salinity, especially in regions where water scarcity, soil degradation, excessive fertilizer application, and harsh weather conditions converge. Mechanisms such as osmotic adjustment, ion regulation, antioxidant defenses, and hormonal and genetic reprogramming are ways that plants use to deal with salinity. In recent times, progress in novel strategies for stress mitigation has been made in omics science, biotechnology, microbiology, and agronomy. This has provided a better understanding of salt stress responses and mitigation in plants. Despite the notable progress made, implementation and research gaps persist in translating laboratory findings into practical field solutions. A comprehensive multidisciplinary approach that integrates thorough real field and multi-location studies, efficient agronomic practices, and supportive policies is needed to solve the problem.

Author details

Kennedy Samuel Banja[1*] and Olusola Jacob Oyetunji[2]

1 Department of Biological Sciences, Taraba State University, Jalingo, Nigeria

2 University of Ibadan, Nigeria

*Address all correspondence to: samuelbanja@tsuniversity.edu.ng

References

[1] Singh A. Soil salinity: A global threat to sustainable development. Soil Use and Management. 2021;**38**:39-67. DOI: 10.1111/sum.12772

[2] Zhang Y, Hou K, Qian H, Gao Y, Fang Y, Xiao S, et al. Characterization of soil salinization and its driving factors in a typical irrigation area of Northwest China. The Science of the Total Environment. 2022;**837**:155808. DOI: 10.1016/j.scitotenv.2022.155808

[3] Tarolli P, Luo J, Park E, Barcaccia G, Masin R. Soil salinization in agriculture: Mitigation and adaptation strategies combining nature-based solutions and bioengineering. iScience. 2024;**2**:27. DOI: 10.1016/j.isci.2024.108830

[4] Hassani A, Azapagic A, Shokri N. Global predictions of primary soil salinization under changing climate in the 21st century. Nature Communications. 2021;**12**:1-7. DOI: 10.1038/s41467-021-26907-3

[5] Daliakopoulos I, Tsanis I, Tsanis I, Koutroulis A, Kourgialas N, Varouchakis A, et al. The threat of soil salinity: A European scale review. The Science of the Total Environment. 2016;**573**:727-739. DOI: 10.1016/j.scitotenv.2016.08.177

[6] Navarro-Torre S, García-Caparrós P, Nogales A, Abreu M, Santos E, Cortinhas A, et al. Sustainable agricultural management of saline soils in arid and semi-arid Mediterranean regions through halophytes, microbial, and soil-based technologies. Environmental and Experimental Botany. 2023;**212**:105397. DOI: 10.1016/j.envexpbot.2023.105397

[7] Haj-Amor Z, Araya T, Kim D, Bouri S, Lee J, Ghiloufi W, et al. Soil salinity and its associated effects on soil microorganisms, greenhouse gas emissions, crop yield, biodiversity and desertification: A review. The Science of the Total Environment. 2022;**843**:156946. DOI: 10.1016/j.scitotenv.2022.156946

[8] Acosta-Motos JR, Ortuño MF, Bernal-Vicente A, Diaz-Vivancos P, Sanchez-Blanco MJ, Hernandez JA. Plant responses to salt stress: adaptive mechanisms. Agronomy. 2017;**7**(1):18. DOI: 10.3389/fpls.2016.00276

[9] Zhao S, Zhang H, Ma C, Wang P, Wu W, Ou X, et al. Regulation of plant responses to salt stress. International Journal of Molecular Science. 2020;**22**(9):4609. DOI: 10.3390/ijms22094609

[10] Flowers TJ, Colmer TD. Plant salt tolerance: Adaptations in halophytes. Annals of Botany. 2015;**115**(3):327-331. DOI: 10.1093/AOB/MCU267

[11] Martins T, Da-Silva C, Shabala S, Striker G, Carvalho I, De Oliveira A, et al. Understanding plant responses to saline waterlogging: insights from halophytes and implications for crop tolerance. Planta. 2023;**259**(1):24. DOI: 10.1007/s00425-023-04275-0

[12] Yuan F, Leng B, Wang B. Progress in studying salt secretion from the salt glands in recretohalophytes: How do plants secrete salt? Frontiers in Plant Science. 2016;7:977. DOI: 10.3389/fpls.2016.00977

[13] Läuchli A, Grattan S. Plant growth and development under salinity stress. In: Jenks MA, Hasegawa PM, Jain SM, editors. Advances in Molecular Breeding toward Drought and Salt Tolerant Crops. Dordrecht: Springer; 2007. DOI: 10.1007/978-1-4020-5578-2_1

[14] Dehnavi A, Zahedi M, Ludwiczak A, Pérez S, Piernik A. Effect of salinity on seed germination and seedling development of sorghum (*Sorghum bicolor* (L.) Moench) genotypes. Agronomy. 2020;**10**:859. DOI: 10.3390/agronomy10060859

[15] Ludwiczak A, Osiak M, Cárdenas-Pérez S, Lubińska-Mielińska S, Piernik A. Osmotic stress or ionic composition: Which affects the early growth of crop species more? Agronomy. 2021;**11**:435. DOI: 10.3390/AGRONOMY11030435

[16] Balasubramaniam T, Shen G, Esmaeili N, Zhang H. Plants' response mechanisms to salinity stress. Plants. 2023;**12**:2253. DOI: 10.3390/plants12122253

[17] Zhang P, Senge M, Dai Y. Effects of salinity stress at different growth stages on tomato growth, yield, and water-use efficiency. Communications in Soil Science and Plant Analysis. 2017;**48**:624-634. DOI: 10.1080/00103624.2016.1269803

[18] Guo J, Lu C, Zhao F, Gao S, Wang B. Improved reproductive growth of euhalophyte Suaeda salsa under salinity is correlated with altered phytohormone biosynthesis and signal transduction. Functional Plant Biology. 2020;**47**:170-183. DOI: 10.1071/FP19215

[19] Van Zandt P, Tobler M, Mouton E, Hasenstein K, Mopper S. Positive and negative consequences of salinity stress for the growth and reproduction of the clonal plant, Iris hexagona. Journal of Ecology. 2003;**91**:837-846. DOI: 10.1046/j.1365-2745.2003.00806.x

[20] Okon O. Effect of salinity on physiological processes in plants. Soil Biology. 2019;**56**:237-262. DOI: 10.1007/978-3-030-18975-4_10

[21] Negrão S, Schmöckel S, Tester M. Evaluating physiological responses of plants to salinity stress. Annals of Botany. 2016;**119**:1-11. DOI: 10.1093/aob/mcw191

[22] Arif Y, Singh P, Siddiqui H, Bajguz A, Hayat S. Salinity induced physiological and biochemical changes in plants: An omic approach towards salt stress tolerance. Plant Physiology and Biochemistry: PPB. 2020;**156**:64-77. DOI: 10.1016/j.plaphy.2020.08.042

[23] Isayenkov S, Maathuis F. Plant salinity stress: Many unanswered questions remain. Frontiers in Plant Science. 2019;**10**:80. DOI: 10.3389/fpls.2019.00080

[24] Shahid M, Sarkhosh A, Khan N, Balal R, Ali S, Rossi L, et al. Insights into the physiological and biochemical impacts of salt stress on plant growth and development. Agronomy. 2020;**10**:938. DOI: 10.3390/agronomy10070938

[25] Hameed A, Ahmed M, Hussain T, Aziz I, Ahmad N, Gul B, et al. Effects of salinity stress on chloroplast structure and function. Cells. 2021;**10**:2023. DOI: 10.3390/cells10082023

[26] Raza A, Tabassum J, Fakhar A, Sharif R, Chen H, Zhang C, et al. Smart reprogramming of plants against salinity stress using modern biotechnological tools. Critical Reviews in Biotechnology. 2022;**43**:1035-1062. DOI: 10.1080/07388551.2022.2093695

[27] Singh P, Choudhary K, Chaudhary N, Gupta S, Sahu M, Tejaswini B, et al. Salt stress resilience in plants mediated through osmolyte accumulation and its crosstalk mechanism with phytohormones. Frontiers in Plant Science. 2022;**13**:1006617. DOI: 10.3389/fpls.2022.1006617

[28] Rivero R, Mestre T, Mittler R, Rubio F, García-Sánchez F, Martínez V.

The combined effect of salinity and heat reveals a specific physiological, biochemical and molecular response in tomato plants. Plant, Cell & Environment. 2014;**37**(5):1059-1073. DOI: 10.1111/pce.12199

[29] Van Zelm E, Zhang Y, Testerink C. Salt tolerance mechanisms of plants. Annual Review of Plant Biology. 2020;**71**:403-433. DOI: 10.1146/annurev-arplant-050718-100005

[30] Almeida D, Oliveira M, Saibo N. Regulation of Na^+ and K^+ homeostasis in plants: towards improved salt stress tolerance in crop plants. Genetics and Molecular Biology. 2017;**40**:326-345. DOI:10.1590/1678-4685-GMB-2016-0106

[31] Yuan F, Wang X, Zhao B, Xu X, Shi M, Leng B, et al. The genome of the recretohalophyte Limonium bicolor provides insights into salt gland development and salinity adaptation during terrestrial evolution. Molecular Plant. 2022;**15**:1024-1044. DOI: 10.1016/j.molp.2022.04.011

[32] Malakar P, Chattopadhyay D. Adaptation of plants to salt stress: the role of the ion transporters. Journal of Plant Biochemistry and Biotechnology. 2021;**30**:668-683. DOI: 10.1007/s13562-021-00741-6

[33] Bohnert H, Shen B. Transformation and compatible solutes. Scientia Horticulturae. 1998;**78**:237-260. DOI: 10.1016/S0304-4238(98)00195-2

[34] Chen H, Jiang J. Osmotic adjustment and plant adaptation to environmental changes related to drought and salinity. Environmental Reviews. 2010;**18**:309-319. DOI: 10.1139/A10-014

[35] Shafi A, Zahoor I, Mushtaq U. Proline accumulation and oxidative stress: Diverse roles and mechanism of tolerance and adaptation under salinity stress. In: Salt Stress, Microbes, and Plant Interactions: Mechanisms and Molecular Approaches. Vol. 2. Singapore: Springer Nature; 2019. pp. 629-300. DOI: 10.1007/978-981-13-8805-7_13

[36] Chen Z, Cuin T, Zhou M, Twomey A, Naidu B, Shabala S. Compatible solute accumulation and stress-mitigating effects in barley genotypes contrasting in their salt tolerance. Journal of Experimental Botany. 2007;**58**(15-16):4245-4255. DOI: 10.1093/jxb/erm284

[37] Melino V, Tester M. Salt-tolerant crops: Time to deliver. Annual Review of Plant Biology. 2022;**74**:671-696. DOI: 10.1146/annurev-arplant-061422-104322

[38] Wang H, Liang L, Liu B, Huang D, Liu S, Liu R, et al. Arbuscular mycorrhizas regulate photosynthetic capacity and antioxidant defense systems to mediate salt tolerance in maize. Plants. 2020;**9**:1430. DOI: 10.3390/plants9111430

[39] Qi J, Luo Y, Huang H, Lu S, Zhao F, Deng Z, et al. Molecular mechanism of response and adaptation of antioxidant enzyme system to salt stress in leaves of Gymnocarpos przewalskii. Plants. 2023;**12**:3370. DOI: 10.3390/plants12193370

[40] Yang Y, Guo Y. Unraveling salt stress signaling in plants. Journal of Integrative Plant Biology. 2018;**60**(9):796-804. DOI: 10.1111/jipb.12689

[41] Zhang X, Sun J, Dong C. Molecular regulations of ethylene signaling in plant salt stress responses. Plant Stress. 2024;**14**:100583. DOI: 10.1016/j.stress.2024.100583

[42] Chen L, Meng Y, Bai Y, Yu H, Qian Y, Zhang D, et al. Starch and

sucrose metabolism and plant hormone signaling pathways play crucial roles in aquilegia salt stress adaptation. International Journal of Molecular Sciences. 2023;**24**:3948. DOI: 10.3390/ijms24043948

[43] Aizaz ML, Jan R, Asaf S, Bilal S, Kim K, Al-Harrasi A. Regulatory dynamics of plant hormones and transcription factors under salt stress. Biology. 2024;**13**:673. DOI: 10.3390/biology13090673

[44] Hassan A, Păcurar A, López-Gresa M, Donat-Torres M, Llinares J, Boscaiu M, et al. Effects of salt stress on three ecologically distinct plantago species. PLoS One. 2016;**11**:e0160236. DOI: 10.1371/journal.pone.0160236

[45] Hassan A, Estrelles E, Soriano P, López-Gresa M, Bellés J, Boscaiu M, et al. Unraveling salt tolerance mechanisms in halophytes: A comparative study on four Mediterranean Limonium species with different geographic distribution patterns. Frontiers in Plant Science. 2017;**8**:277135. DOI: 10.3389/fpls.2017.01438

[46] Alnusairi G, Mazrou Y, Qari S, Elkelish A, Soliman M, Eweis M, et al. Exogenous nitric oxide reinforces photosynthetic efficiency, osmolyte, mineral uptake, antioxidant, expression of stress-responsive genes, and ameliorates the effects of salinity stress in wheat. Plants. 2021;**10**:1693. DOI: 10.3390/plants10081693

[47] Latef A, Tahjib-Ul-Arif M, Rhaman M. Exogenous auxin-mediated salt stress alleviation in Faba bean (*Vicia faba* L.). Agronomy. 2021;**11**:547. DOI: 10.3390/AGRONOMY11030547

[48] Sadak M, Sękara A, Al-Ashkar I, Habib-Ur-Rahman M, Skalický M, Brestič M, et al. Exogenous aspartic acid alleviates salt stress-induced decline in growth by enhancing antioxidants and compatible solutes while reducing reactive oxygen species in wheat. Frontiers in Plant Science. 2022;**13**:987641. DOI: 10.3389/fpls.2022.987641

[49] Islam M, Jahan K, Sen A, Urmi T, Haque M, Ali H, et al. Exogenous application of calcium ameliorates salinity stress tolerance of tomato (*Solanum lycopersicum* L.) and enhances fruit quality. Antioxidants. 2023;**12**:558. DOI: 10.3390/antiox12030558

[50] Sá F, Brito M, Silva L, Moreira R, Paiva E, Souto L. Exogenous application of phytohormones mitigates the effect of salt stress on *Carica papaya* plants. Revista Brasileira de Engenharia Agrícola e Ambiental. 2020;**24**:170-175. DOI: 10.1590/1807-1929/agriambi.v24n3p170-175

[51] Sardar H, Khalid Z, Ahsan M, Naz S, Nawaz A, Ahmad R, et al. Enhancement of salinity stress tolerance in lettuce (*Lactuca sativa* L.) via foliar application of nitric oxide. Plants. 2023;**12**:1115. DOI: 10.3390/plants12051115

[52] Linić I, Mlinarić S, Brkljačić L, Pavlović I, Smolko A, Salopek-Sondi B. Ferulic acid and salicylic acid foliar treatments reduce short-term salt stress in Chinese cabbage by increasing phenolic compounds accumulation and photosynthetic performance. Plants. 2021;**10**:2346. DOI: 10.3390/plants10112346

[53] Oyetunji OJ, Banja SK. Influence of arbuscular mycorrhizal fungi (glomus mosseae (T. H. Nicolson & Gerd) Gerd. & Trappe) on some physiological responses of cassava (Manihot esculenta Crantz) to drought and salinity. Journal of Food, Agriculture & Environment. 2020;**18**(1):51-57

[54] Ali B, Hafeez A, Ahmad S, Javed MS, Afridi M, Dawoud T, et al. *Bacillus thuringiensis* PM25 ameliorates oxidative damage of salinity stress in maize via regulating growth, leaf pigments, antioxidant defense system, and stress-responsive gene expression. Frontiers in Plant Science. 2022;**13**:921668. DOI: 10.3389/fpls.2022.921668

[55] Hu G, Zhao Z, Wei Y, Hu J, Zhou Y, Li J, et al. *Trichoderma asperellum* 22043: Inoculation promotes salt tolerance of tomato seedlings through activating the antioxidant system and regulating stress-resistant genes. Journal of Fungi. 2025;**11**:253. DOI: 10.3390/jof11040253

[56] Talaat N, Todorova D. Antioxidant machinery and glyoxalase system regulation confers salt stress tolerance to wheat (*Triticum aestivum* L.) plants treated with melatonin and salicylic acid. Journal of Soil Science and Plant Nutrition. 2022;**22**:3527-3540. DOI: 10.1007/s42729-022-00907-8

[57] Gamalero E, Glick B. Recent advances in bacterial amelioration of plant drought and salt stress. Biology. 2022;**11**:437. DOI: 10.3390/biology11030437

[58] Gohari G, Panahirad S, Sepehri N, Akbari A, Zahedi S, Jafari H, et al. Enhanced tolerance to salinity stress in grapevine plants through application of carbon quantum dots functionalized by proline. Environmental Science and Pollution Research. 2021;**28**:42877-42890. DOI: 10.1007/s11356-021-13794-w

[59] Khan Z, Gul H, Rauf M, Arif M, Hamayun M, Sajid ZA, et al. *Sargassum wightii* aqueous extract improved salt stress tolerance in *Abelmoschus esculentus* by mediating metabolic and ionic rebalance. Frontiers in Marine Science. 2022;**9**:853272. DOI: 10.3389/fmars.2022.853272

[60] Wang T, Xun H, Wang W, Ding X, Tian H, Hussain S, et al. Mutation of GmAITR genes by CRISPR/Cas9 genome editing results in enhanced salinity stress tolerance in soybean. Frontiers in Plant Science. 2021;**12**:779598. DOI: 10.3389/fpls.2021.779598

[61] Tran M, Son G, Song Y, Nguyen N, Park S, Thach T, et al. CRISPR-Cas9-based precise engineering of SlHyPRP1 protein towards multi-stress tolerance in tomato. Frontiers in Plant Science. 2023;**14**:1186932. DOI: 10.3389/fpls.2023.1186932

[62] Farhat S, Jain N, Singh N, Sreevathsa R, Dash P, Rai R, et al. CRISPR-Cas9 directed genome engineering for enhancing salt stress tolerance in rice. Seminars in Cell & Developmental Biology. 2019;**96**:91-99. DOI: 10.1016/j.semcdb.2019.05.003

[63] Nazir R, Mandal S, Mitra S, Ghorai M, Das N, Jha N, et al. CRISPR/Cas genome-editing toolkit to enhance salt stress tolerance in rice and wheat. Physiologia Plantarum. 2022;**174**(2):e13642. DOI: 10.1111/ppl.13642

[64] Ganie S, Molla K, Henry R, Bhat K, Mondal T, Mondal T. Advances in understanding salt tolerance in rice. Theoretical and Applied Genetics. 2019;**132**:851-870. DOI: 10.1007/s00122-019-03301-8

[65] Hussain TM, Chandrasekhar T, Hazara M, Sultan Z, Saleh BK, Gopal GR. Recent advances in salt stress biology – a review. Biotechnology and Molecular Biology Review. 2008;**3**:8-13

[66] Morton M, Awlia M, Al-Tamimi N, Saade S, Pailles Y, Negrão S, et al. Salt stress under the scalpel – dissecting the genetics of salt tolerance. The Plant Journal. 2018;**97**:148-163. DOI: 10.1111/tpj.14189

[67] Pan LJ, Li J, Yin B, Fu C. Advances of salt stress-responsive transcription factors in plants. Sheng wu gong cheng xue bao = Chinese journal of biotechnology. 2022;**38**(1):50-65. DOI: 10.13345/j.cjb.210135

[68] Pradhan S, Bhattacharjee A, Dubey S, Sharma S. Ameliorating salt stress in tomato by a top-down approach of acclimatizing the rhizosphere microbiome. Physiologia Plantarum. 2025;**177**:e770071. DOI: 10.1111/ppl.70071

[69] Liang X, Li J, Yang Y, Jiang C, Guo Y. Designing salt stress-resilient crops: Current progress and future challenges. Journal of Integrative Plant Biology. 2023;**66**:303-329. DOI: 10.1111/jipb.13599

[70] Hasan M, Rahman M, Corpas F, Rahman M, Jahan M, Liu X, et al. Salt stress tolerance in rice (*Oryza sativa* L.): A proteomic overview of recent advances and future prospects. Plant Stress. 2023;**11**:100307. DOI: 10.1016/j.stress.2023.100307

[71] Mann A, Lata C, Kumar N, Kumar A, Kumar A, Sheoran P. Halophytes as new model plant species for salt tolerance strategies. Frontiers in Plant Science. 2023;**14**:1137211. DOI: 10.3389/fpls.2023.1137211

[72] Wang C, Han G, Yang Z, Li Y, Wang B. Plant salinity sensors: Current understanding and future directions. Frontiers in Plant Science. 2022;**13**:859224. DOI: 10.3389/fpls.2022.859224

[73] Park H, Kim W, Yun D. A new insight of salt stress signaling in plant. Molecules and Cells. 2016;**39**:447-459. DOI: 10.14348/molcells.2016.0083

[74] Tibesigwa D, Zhuang W, Matola S, Zhao H, Li W, Yang L, et al. Molecular insights into salt stress adaptation in plants. Plant. 2025;**48**:5604-5615. DOI: 10.1111/pce.15544

[75] Evelin H, Devi T, Gupta S, Kapoor R. Mitigation of salinity stress in plants by Arbuscular mycorrhizal symbiosis: Current understanding and new challenges. Frontiers in Plant Science. 2019;**10**:450967. DOI: 10.3389/fpls.2019.00470

Chapter 5

Abiotic Stress in Plants: Biosurfactants as Tools to Enhance Plant Resilience

Bruna Sthefanie Paz de Souza, Glacielen Ribeiro de Souza, Analyse Villanueva Gaete, Gabriela Petroceli Mota, João Vitor Pereira de Souza, Daiane Carvalho Baía, Marília Berbert Molina and Aline Chaves Intorne

Abstract

With the intensification of agricultural activities, there has been a persistent release of pollutants into the environment such as metals, hydrocarbons, and pesticides, resulting in the degradation of natural cycles and the imposition of multiple abiotic stresses on plants. These conditions significantly compromise plant growth, productivity, and overall balance. In this context, microbial biosurfactants emerge as promising multifunctional agents, acting both in the bioremediation of contaminated environments and in promoting plant tolerance to adverse conditions. Produced by a wide variety of microorganisms, including bacteria and fungi, these amphiphilic molecules reduce surface and interfacial tension, enhancing the solubilization, mobilization, and bioavailability of contaminants and nutrients in complex ecosystems. These physicochemical properties confer a strategic role to biosurfactants in mitigating the phytotoxic effects caused by excessive levels of metals, persistent organic compounds, and high salinity, promoting a more favorable rhizospheric environment for plant development. Furthermore, recent evidence shows that biosurfactants can alter the structure and activity of plant-associated microbiota, stimulating colonization by plant growth-promoting microorganisms and participating in cellular signaling with the induction of defense responses. Such responses include the regulation of metabolic pathways associated with osmotic balance, antioxidant activity, and the expression of stress-responsive genes. Thus, the application of microbial biosurfactants represents an innovative, sustainable, and biologically integrated strategy to enhance plant resilience against abiotic stress conditions, while contributing to the rehabilitation of degraded environments. Advances in microbial biotechnology have enabled the optimization of production and structural characterization of these compounds, consolidating their applicability in sustainable agricultural and environmental systems, which will be explored in this chapter.

Keywords: microbial surfactants, biodegradation, abiotic stress, plant physiology, Environmental biotechnology

IntechOpen

1. Introduction

The intensification of agricultural activities has contributed to soil contamination by pollutants such as herbicides, hydrocarbons, and metals, representing one of the main causes of environmental degradation [1]. This problem is further aggravated by other abiotic factors, such as soil salinization and acidification, prolonged drought periods, and abrupt temperature fluctuations [2]. The combined effect of these conditions, along with increased environmental pollution and the negative impacts of climate change, poses significant challenges to modern agroecosystems [3, 4]. Abiotic stresses compromise soil resilience and affect photosynthesis, nutrient uptake, and plant development, leading to reduced agricultural productivity [5].

This scenario underscores the urgent need for sustainable strategies capable of mitigating the effects of environmental degradation and promoting the recovery of affected ecosystems. Microbial biotechnology, which employs microorganisms to produce valuable products on a large scale, has emerged as a promising approach for this purpose, offering efficient and ecological solutions for the remediation of contaminated environments [6]. Microorganisms have the capacity to degrade numerous toxic compounds and thus can contribute to restoring environmental balance and promoting the sustainability of impacted areas [7, 8].

Among the diverse metabolites produced by microorganisms, biosurfactants have attracted attention due to their multifunctional properties. These amphiphilic biomolecules can reduce surface tension between liquids and solids, favoring the solubilization and biodegradation of hydrophobic substances such as hydrocarbons, metals, and some agrochemicals [9].

Beyond their direct application in bioremediation processes, recent studies have revealed the role of biosurfactants in promoting plant resilience against abiotic stress [10, 11]. These bioactive molecules act in a complex manner within the rhizosphere, modulating rhizospheric microbiota, increasing nutrient availability, inducing antioxidant responses, and contributing to osmotic balance and ionic homeostasis in plants. These mechanisms are fundamental for plant survival under adverse conditions such as salinity, drought, oxidative stress, and the presence of toxic metals [12].

Biosurfactants are produced by various microorganisms, especially bacteria and fungi, and their synthesis is influenced by factors such as substrate type, physicochemical conditions of the medium, and metabolic characteristics of the strains used [13]. However, these factors act interdependently, modulating the expression of genes involved in biosynthesis and influencing the type and quantity of biosurfactants produced. Genera such as *Pseudomonas* and *Aspergillus* have stood out for their high production capacity even in contaminated environments, suggesting an adaptive function of these molecules in response to stress [14, 15]. In this context, production can be interpreted not only as a metabolic phenomenon but also as part of microbial strategies for survival and detoxification, reinforcing their potential for applications in environments impacted by xenobiotics [16].

Compared to synthetic surfactants, biosurfactants present significant advantages, including greater biodegradability, lower toxicity, and environmental compatibility, making them safe and effective alternatives for decontaminating areas impacted by chemical compounds [17, 18].

By understanding the ecological, physiological, and molecular mechanisms through which biosurfactants promote plant tolerance to adverse conditions, their broad applicability as a strategic tool to address challenges imposed by abiotic stresses becomes evident. To this end, this chapter will begin with a detailed approach to the main types of abiotic stresses affecting agricultural systems, their causes, impacts, and implications for plant productivity. Subsequently, we will delve into the characteristics and potential of biosurfactants.

2. Abiotic stresses in agricultural systems: Causes, impacts, and challenges

Abiotic stresses are non-biological environmental factors that negatively affect crop growth, productivity, and quality [19]. These stresses have gained increasing attention due to their significant impact on global food security, with estimates indicating that approximately 96.5% of the world's agricultural lands are affected, potentially causing yield losses of up to 60% [20]. The intensification of human activities since the Industrial Revolution, particularly the extensive use of chemical inputs and rapid industrialization, has contributed to soil degradation, environmental pollution, and disruption of agricultural ecosystems [21, 22].

Among the most critical abiotic stressors are emerging contaminants, a diverse group of artificial and natural chemical substances whose environmental and health impacts are not yet fully regulated or monitored [23, 24]. Agricultural practices, especially the global use of fertilizers and pesticides, are major contributors. The production of chemical fertilizers exceeded 220 million tons in 2019 [25], and phosphate fertilizers are a notable source of toxic metals such as cadmium (Cd), lead (Pb), zinc (Zn), nickel (Ni), and copper (Cu), which accumulate in soils and plants over time, compromising soil fertility and food safety [25].

Pesticides, which account for roughly one-third of global agricultural production, are also major stressors, spreading widely across soil, water, air, and plant tissues. Herbicides represent 47.5% of the total, followed by insecticides (29.5%), fungicides (17.5%), and other less-used agrochemicals (5.5%) [26]. These compounds can persist in the environment, bioaccumulate within organisms, and biomagnify along food chains, affecting non-target species and ecosystem balance [27–29].

Climate change acts synergistically with these anthropogenic factors, intensifying abiotic stresses such as drought, salinity, and extreme temperatures. According to the Intergovernmental Panel on Climate Change [30, 31], there is a 66% probability that the average global temperature will rise above 1.5°C between 2030 and 2052. In more severe scenarios, the increase may surpass 2°C by 2050, exacerbating environmental pressures on agricultural systems [20]. Together, chemical contamination, environmental degradation, and climate change create a multifaceted challenge for global agriculture, demanding the development of resilient crop varieties, sustainable management practices, and innovative mitigation strategies.

2.1 Main types of abiotic stress affecting agricultural crops

Given the severity of the consequences of abiotic stresses on agroecosystems and the need to seek sustainable resilience solutions, it is essential to understand how these factors affect plant development and impact agricultural activities. The following section briefly discusses the main abiotic stresses and their effects on agricultural crops.

2.1.1 High temperatures

Temperature is a critical factor in biological systems, as it regulates chemical reaction rates, which increase up to an optimum point before sharply declining [32].

In plants, heat stress occurs when ambient temperatures exceed the ideal range for growth, potentially causing severe physiological and cellular damage [20]. Moderate heat can trigger morphological, physiological, and biochemical adjustments that contribute to acclimation, while excessively high temperatures can impair development and even cause plant death. Crops such as rice, wheat, soybean, maize, cotton, sorghum, and tomato are particularly sensitive [33].

High temperatures reduce photosynthetic efficiency by affecting photosystem II, electron transport, and the proportion of active reaction centers, which directly impacts productivity [34, 35]. Reproductive processes, especially early seed and fruit development, are also affected, leading to lower seed filling and reduced fruit quality [36, 37]. Elevated nighttime temperatures exacerbate these effects by increasing respiration, consuming photoassimilates, and reducing carbon available for growth [8].

Heat stress often occurs in combination with drought, and their synergistic effects further reduce crop yields and agricultural productivity, contributing to higher food prices [38].

2.1.2 Drought

Drought is characterized by prolonged precipitation deficits, causing significant soil water imbalance [30]. Rising global temperatures and more frequent extreme climate events, such as heatwaves and long dry spells, have intensified drought occurrences worldwide, with projections indicating increasing frequency and severity [20, 30].

Water scarcity is a major challenge for agriculture, which currently consumes 70% of global freshwater [7]. Drought affects multiple physiological and metabolic processes, reducing plant growth and productivity. Its impact depends on stress intensity, duration, and the plant's phenological stage, with the reproductive phase being especially sensitive, impairing grain filling, flower formation, seed viability, and longevity [20, 39]. Seed germination is also affected in major crops like maize, sorghum, and wheat [40].

Early vegetative symptoms include leaf wilting, reduced height, and delayed bud and flower development. Reduced leaf area limits photosynthesis, while stomatal closure to conserve water restricts CO_2 intake and promotes ROS accumulation, damaging cellular structures and the photosynthetic apparatus [19, 40]. Drought also reduces water and nutrient uptake, decreases cell turgor, and shifts resources toward osmoprotectant production, ultimately limiting growth and crop yield [20].

2.1.3 Cold

Climate change has intensified cold stress, increasing the frequency of low-temperature events, snowstorms, and abrupt temperature fluctuations in major agricultural regions [41]. About 64% of global land experiences minimum temperatures below 0°C, making cold stress highly relevant to crops and ecosystems [42].

Sensitivity to low temperatures varies by species, cultivar, and phenological stage, with plants classified as sensitive, cold-tolerant, or freezing-tolerant. Although some develop cold acclimation, abrupt temperature changes often compromise this adaptation [20, 43]. Cold stress reduces growth and photosynthesis, increases membrane rigidity, alters lipid-to-protein ratios, impairs water absorption, and triggers ROS accumulation and metabolic dysfunctions in crops such as wheat and maize [41].

Even non-lethal low temperatures negatively affect vegetative and reproductive development, causing chlorosis, wilting, floral abortion, poor germination, and reduced yield [42, 44]. Freezing conditions exacerbate damage via ice crystal formation, dehydration, cell rupture, protein denaturation, and impaired seed filling [45, 46]. In perennials, winter survival depends on the balance between deacclimation and reacclimation, processes highly influenced by temperature fluctuations [47, 48].

Cellular water relations and soluble sugars contribute to cold tolerance by stabilizing structures, regulating osmotic balance, and preventing excessive dehydration [42, 49]. Additionally, excess soil salts from agrochemical applications represent a related abiotic stress, further impacting plant physiology.

2.1.4 Salinity

Saline soil is defined as soil with electrical conductivity (EC) above 4 dS m^{-1} in the root zone, corresponding to 40 mM NaCl, which reduces the yield of most cultivated plants [50]. High concentrations of soluble salts, particularly Na^+ and Cl^-, impair physiological and biochemical processes, reducing biomass and crop productivity [51]. Salinity affects photosynthesis, relative water content, leaf CO_2 uptake, induces stomatal closure, pollen sterility, and membrane degradation, compromising quality and yield [52].

A crop's tolerance depends on its ability to extract water and nutrients while limiting toxic ion accumulation [51]. Salt accumulation in leaves accelerates senescence, disrupts nutrient uptake, interferes with photosynthesis, reduces gas exchange, promotes chlorophyll degradation, and damages chloroplasts, triggering defense mechanisms such as vacuolar ion compartmentalization and ROS detoxification [20, 52]. Stress is most severe during sensitive stages like germination, seedling establishment, and flowering [38, 53].

Salt stress can be divided into three phases: homeostasis (normal growth), stress (activation of defense genes), and suffering (metabolic slowdown and possible cell death) [50]. Tolerant species respond faster, producing osmolytes, antioxidant enzymes, and heat shock proteins, whereas susceptible plants exhibit slower, less efficient responses [52].

2.1.5 Metals

Metals can be classified as essential, such as copper (Cu), zinc (Zn), and iron (Fe), which perform biological functions, and non-essential, like cadmium (Cd), mercury (Hg), and lead (Pb), which are toxic even at low concentrations [54, 55]. Anthropogenic activities—including fertilizer and pesticide use, industrial and domestic waste disposal, irrigation with contaminated water, mining, and emissions from power plants—have led to persistent metal accumulation in soils and leaching into groundwater, increasing environmental risks [25, 55–57].

Metals are non-biodegradable. Some are mobile and absorbed by roots via diffusion or transporters, while others accumulate in the root zone [58, 59]. Excess metals disrupt plant physiological, biochemical, and morphological functions by replacing essential ions, denaturing enzymes, damaging membranes, and impairing photosynthesis, respiration, and homeostasis [56, 58]. Metal accumulation also stimulates ROS production, causing lipid peroxidation, DNA and protein damage, and cell death [19, 56].

Visible symptoms include stunted growth, chlorosis, root darkening, leaf drop, and reduced seed germination. Contaminated soils often show reduced vegetation, favoring erosion and contaminant dispersion, and heavy metals negatively impact soil microbial diversity and activity, affecting soil fertility and ecosystem balance [60].

2.1.6 Pesticides

Pesticides are chemical compounds used to control harmful organisms in agriculture, including weeds, insects, and fungi [61]. While effective for pest management, excessive use represents a significant abiotic stress in agroecosystems [27].

Many synthetic pesticides, such as chlordane and thiobencarb, are persistent organic pollutants (POPs) with low degradation rates, high chemical stability, and bioaccumulation potential [62, 63]. They can disperse to non-target environments through leaching, volatilization, runoff, or spray drift, and their fate depends on physicochemical properties and environmental conditions [23, 64].

Persistent pesticide residues negatively affect soil microbiota, reducing microbial diversity and impairing biochemical processes essential for plant growth [23, 62]. In plants, they induce oxidative stress, DNA damage, chlorosis, necrosis, inhibited photosynthesis, and, in severe cases, cell death [26, 27]. Herbicides can also harm non-target plant species [65].

Pesticides impact beneficial organisms such as pollinators and nitrogen-fixing bacteria, for example, Roundup® affecting Rhizobium nodules [66, 67]. Continuous use exerts selective pressure, promoting resistance in target populations, such as *Amaranthus palmeri* developing multi-herbicide resistance, which reduces efficacy and increases environmental impacts [68, 69].

Given the multiple negative effects caused by abiotic stresses in agricultural systems, the need for sustainable strategies that mitigate these impacts and strengthen agroecosystem resilience becomes evident.

3. Microbial biosurfactants: A promising option to mitigate abiotic stresses

3.1 Structural characteristics and functional properties of biosurfactants

Biosurfactants are surface-active molecules of biological origin, primarily produced by microorganisms such as bacteria, yeasts, and fungi [70]. These compounds have an amphiphilic nature, simultaneously possessing a hydrophilic (polar) portion, which may include groups such as carbohydrates, amino acids, peptides, phosphates, or carboxylic acids, and a hydrophobic (non-polar) portion, generally composed of saturated or unsaturated fatty acids, which can be linear or branched (**Figure 1**) [70, 71]. This structural characteristic allows biosurfactants to accumulate at interfaces between immiscible liquids, such as water and oil, reducing surface tension (liquid-air) and interfacial tension (liquid-liquid), thereby promoting the emulsification and dispersion of hydrophobic compounds in the aqueous medium [11].

At low concentrations, they exist as monomers and progressively reduce surface tension. Above the critical micelle concentration (CMC), the molecules self-organize into micelles [72]. In polar media (e.g., water), conventional micelles form, with hydrophilic heads, oriented outward toward the surrounding medium and hydrophobic tails oriented inward. In non-polar media (e.g., oils), reverse micelles form, with hydrophobic tails facing outward and hydrophilic heads organized inside the micelle.

Unlike synthetic surfactants derived from petroleum, microbial surfactants stand out for being biodegradable, exhibiting low toxicity, greater environmental compatibility, and broad stability across variations in pH and temperature. Additionally, they are produced under moderate cultivation conditions [73]. The structural versatility of

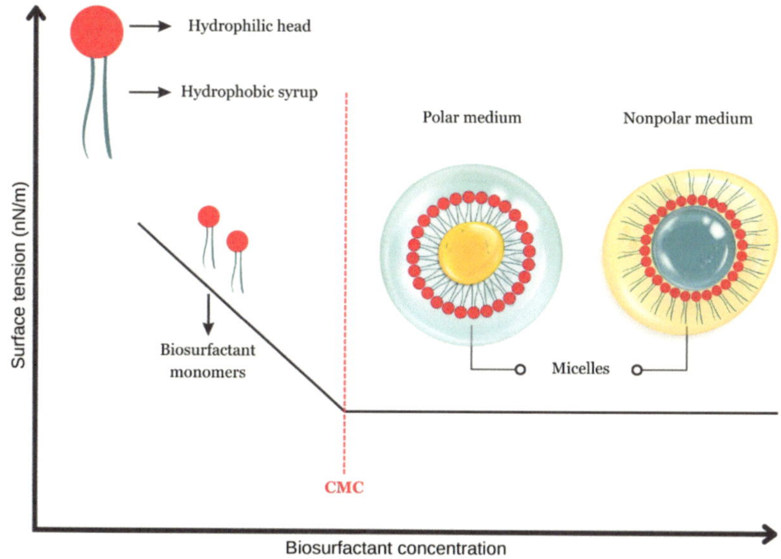

Figure 1.
Schematic representation of biosurfactant self-organization and micelle formation according to the polarity of the medium, highlighting the critical micelle concentration (CMC) as the limit above which biosurfactant molecules aggregate into micelles.

these compounds also enables a diverse range of applications, notably in sectors such as agriculture, cosmetics, pharmaceuticals, food, and bioremediation processes [11]. Beyond their role as surfactant agents, biosurfactants perform important biological functions, including antimicrobial, antiviral, insecticidal, and anti-adhesive activities, which are attributed to their ability to interact with cell membranes, alter cell wall permeability, inhibit key enzymes, and disrupt vital processes such as DNA replication [74]. These properties make biosurfactants promising tools for managing phytopathogens, as well as acting as bio-stimulants for plant growth and mitigators of abiotic stresses such as salinity, metals, and pesticides [75].

The efficiency of biosurfactants in reducing interfacial tension is directly linked to their chemical composition. According to molecular weight, these compounds can be grouped into two categories: low molecular weight biosurfactants, such as glycolipids and lipopeptides, and high molecular weight biosurfactants, including lipopolysaccharides, polymeric surfactants, and lipoproteins (**Table 1**) [71, 79]. Generally, low molecular weight biosurfactants are more effective at reducing surface tension, while high molecular weight biosurfactants excel in stabilizing emulsions [80]. Biosurfactant classification can also consider their microbial origin or chemical composition, including fatty acids, glycolipids, peptides, lipopeptides, phospholipids, and compounds with antibiotic activity [81]. The chemical particularities of these molecular structures directly influence their biological functions and potential applications [71].

The efficacy of a biosurfactant is often evaluated through the Critical Micelle Concentration (CMC), a parameter that represents the lowest concentration required for micelle formation and maximum surface tension reduction. Efficient biosurfactants can reduce the surface tension of water from approximately 72 mN/m (millinewtons per meter) to values between 30 and 35 mN/m, while the interfacial tension between oil and water can reach values as low as 1 mN/m. Compared to synthetic

Class	Subtypes	Producers	Main characteristics	Applications	References
Glycolipids	Trehalolipids, rhamnolipids, mannosylerythritol lipids (MELs), sophorolipids	*Pseudomonas* spp., *Starmerella bombicola*, *Moesziomyces* sp., *Rhodococcus* spp., *Arthrobacter* spp., *Candida* spp.	Composed of carbohydrate + fatty acid; effective at low concentrations; excellent surface tension reducers	Cosmetics, pharmaceuticals, antimicrobials, bioremediation, food industry	Dini et al. [39]; Faccioli et al. [76]
Lipopeptides	Surfactin, iturin, fengycin, liquenisin, viscosin, amphisin, tolaasin, syringomycin	*Bacillus* spp., *Fusarium* sp., *Pseudomonas* spp.	Cyclic peptide linked to fatty acid; antimicrobial and hemolytic activity; good physicochemical stability	Biocontrol, antimicrobials, human health, environmental biotechnology, food industry	Faccioli et al. [76]; da Silva et al. [77]
Phospholipids	Phosphatidylcholine, phosphatidylethanolamine	*Acinetobacter* spp., *Thiobacillus thiooxidans*	Phosphate head + fatty acid tail; production enhanced by hydrocarbons; affect membrane fluidity	Bioremediation, micellar systems	Dini et al. [39]; Faccioli et al. [76]
Polymeric surfactants	Emulsan, alasan, liposan	*Acinetobacter calcoaceticus*, *A. radioresistens*, *Candida albicans*, *Yarrowia lipolytica*	High molecular weight (protein + polysaccharide); stable emulsion-forming agents; high viscosity	Emulsion stabilization, petroleum industry, food sector	Srivastava et al. [2]; Bhadra et al. [78]
Glycolipoproteins	—	*Bacillus subtilis*, *Streptomyces* sp., *Mycobacterium* sp., *Lactobacillus acidophilus*, *L. jensenii*, *Acinetobacter* spp.	Lipid chains linked to proteins; structural or antimicrobial function	Biocontrol, antimicrobials, biodegradation	Dini et al. [39]
Particulate biosurfactants	Vesicles, whole microbial cells	*Acinetobacter* sp., *Mycobacterium* sp., *Bacillus* sp., *Staphylococcus aureus*	Cell surfaces or extracellular structures with biosurfactant properties; insoluble; capable of hydrocarbon adsorption adsorção de hidrocarbonetos	Bioremediation	Srivastava et al. [2]; Bhadra et al. [78]

Table 1.
Classification of biosurfactants based on chemical structure, microbial producers, main characteristics, and applications.

surfactants, biosurfactants exhibit significantly lower CMCs, ranging between 0.15 and 30 mg L^{-1}, which demonstrates their superiority in terms of performance and environmental sustainability [82].

Among the most studied biosurfactants are rhamnolipids, surfactin, mannosylerythritol lipids (MELs), and sophorolipids, due to their emulsification efficacy and applicability in the pharmaceutical, cosmetic, food, and agricultural industries. These molecules have complex structures, often composed of mixtures of congeners, isomers, and homologs. This complexity requires the use of high-precision analytical techniques for characterization, such as liquid chromatography coupled to mass spectrometry (LC-MS), gas chromatography coupled to mass spectrometry (GC-MS), mass spectrometry (MS), Fourier-transform infrared spectroscopy (FTIR), and nuclear magnetic resonance (NMR) [74].

3.2 Microorganism producers and production process aspects

Various species of bacteria, yeasts, and fungi isolated from different habitats have demonstrated promising potential for biosurfactant production, exhibiting enormous structural and functional diversity [83]. Among bacteria, certain members of the genera *Pseudomonas*, *Actinobacter*, *Sphingomonas*, *Halomonas*, *Bacillus*, *Mycobacterium*, *Alcanivorax*, *Rhodococcus*, *Pseudoalteromonas*, and *Arthrobacter* show remarkable capability for producing these compounds [84]. The species *Pseudomonas aeruginosa*, *Bacillus subtilis*, and *Acinetobacter calcoaceticus*, recognized for their high metabolic versatility, stand out due to their ability to utilize various natural substrate sources and synthesize different types of biosurfactants [82, 83]. *Pseudomonas aeruginosa* is the main producer of rhamnolipids, capable of producing up to 100 g L^{-1} of the compound, making the production process highly economical compared to many synthetic surfactants [85, 86].

Although less explored, fungal-origin biosurfactants are also relevant, accounting for about 20% of the total production of these metabolites [87, 88]. Generally, these biosurfactants exhibit remarkable chemical structural diversity, with some produced exclusively by fungi, such as sophorolipids, mannosylerythritol lipids, cellobiose lipids, xylolipids, lipid polyols, and hydrophobins [2, 89]. This versatility is an important feature as it can result in compounds with distinct properties and greater application potential [87]. The main producers include the yeasts *Candida lipolytica*, *Candida ishiwadae*, *Candida bombicola*, and *Yarrowia lipolytica* [90], as well as fungi from the genera *Trichoderma*, *Fusarium*, *Aspergillus*, and *Penicillium* [81, 88].

Recent bioprospecting studies, including exploration of marine environments and extreme ecosystems, have revealed new biosurfactant-producing microorganisms with robust surfactant properties [84, 91]. Adapted to high salinity and pressure, marine microorganisms act as natural biosurfactant producers, offering promising perspectives for industrial applications [92]. The isolation of strains such as *Yarrowia lipolytica* LMS 24B, *Exiguobacterium* sp. R58, and *Marinobacter* sp. MCTG107b highlights the potential of these habitats for this purpose [60, 93, 94]. Extreme environments also emerge as important reservoirs of biosurfactants with biotechnologically relevant properties. Extremophilic microorganisms, thriving under extreme temperature, pH, or salinity conditions, inherently possess the ability to produce these compounds [95, 96]. For instance, the Antarctic strain *Moesziomyces antarcticus* secretes mannosylerythritol lipid, a biosurfactant widely used in the cosmetic industry [97]. Another example is *Bacillus alveayuensis*, a thermo-halophilic bacterium isolated from an oil reservoir, which produces a mixture of surfactins, iturins, and fengycins [98].

The industrial application of surfactant molecules has expanded in recent years, and prospects for large-scale production growth are encouraging [99]. The global biosurfactant market value was USD 4.2 billion in 2017, with an expected annual growth rate of 5.6%. However, optimized techniques are required to ensure the generation of products with high purity and quality, as well as yields and productivities compatible with commercial-scale production [100]. Besides the adequate selection of microbial species or strains and their origin, the synthesis of these biomolecules is strongly influenced by several environmental and operational factors, especially fermentation technology, physicochemical parameters affecting microbial metabolism, and the nutritional characteristics of the cultivation medium [101, 102].

The definition of carbon and nitrogen sources, minerals, and the carbon/nitrogen ratio used in microbial cultivation is highly relevant for biosurfactant synthesis, being crucial for achieving high product concentrations and robust yields [103, 104], as well as for the composition of the generated metabolite [105]. Different types of carbon sources have been evaluated for the production process [106], with results indicating that both the concentration and nature of the substrate influence key fermentation parameters [103]. Common sources include carbohydrates, vegetable oils, animal fats, starchy materials, and hydrocarbons [104], but various other residues have been investigated as low-cost options for the process, such as petroleum effluents, molasses, glycerol, used frying oil, and agro-industrial effluents [107–109]. Besides decisively influencing the yield and productivity of the fermentation process, the carbon source can represent up to 50% of total production costs, making its choice a crucial factor for economic feasibility [80].

Nitrogen also plays a determining role in surfactant production, being essential for microbial growth and bioactive metabolite synthesis [75]. Various organic and inorganic nitrogen sources have been used in fermentation, including yeast extract, nitrates, ammonium salts, and urea, with variable results depending on the strain used [88, 103]. The carbon/nitrogen ratio is also a variable with great influence on the process, as it affects microbial biomass formation and biosurfactant accumulation. Lower nitrogen concentrations (i.e., higher C/N ratios) inhibit bacterial growth, directing cellular metabolism toward the synthesis of these biomolecules [90].

Besides carbon and nitrogen, biosurfactant synthesis depends on the presence of minerals in the cultivation medium. Metal ions such as potassium, magnesium, iron, manganese, cobalt, and zinc, involved in various metabolic processes, have been described as fundamental for production [103]. The salinity of the medium is also an influential factor in biosurfactant production, but tolerance limits depend on the species used in fermentation. NaCl concentration has been shown to be critical for *Serratia marcescens* [110] and *Vibrio* spp. [111].

To establish operational conditions that enable viable biosurfactant production, it is also essential to define the optimal values of variables that most impact microbial growth and target product formation, such as pH, temperature, and oxygen transfer. Generally, the optimal fermentation temperature is between 25 and 30°C, and pH ranges from 7.0 to 8.0 [13, 75, 105], which can also affect the final metabolite composition [103]. The dissolved oxygen concentration in the fermentation medium is also relevant, requiring adequate agitation and aeration to maintain metabolic activity and microbial production [109]. Submerged fermentation, where microbial cultivation occurs in a liquid medium and metabolites are released into the broth, is the most widely used technique for production [112] and has been highly effective for yeast and bacterial biosurfactant producers [84]. Solid-state fermentation, involving microbial growth on solid materials without a free liquid phase, is gaining popularity as a potential strategy for biosurfactant production, particularly because it allows the use of solid agro-industrial residues in the cultivation medium formulation [109, 113] and helps circumvent the common foaming problem in submerged fermentations [84].

Regarding fermentation process mode, batch systems have been the most employed, demonstrating robust results for different species, substrates, and biosurfactant types [1, 77]. Studies using fed-batch systems have shown improved biosurfactant production [114] and productivity [115] with various microbial species and culture media. The use of continuous systems is more limited but cited as a promising strategy for the process [7, 36].

At the end of the production process, the fermented broth is directed to downstream operations, involving product separation and purification—crucial activities for economic viability at large scale [109]. Cell separation is generally done by centrifugation, and the surfactant-containing supernatant undergoes purification steps, including acidification and extraction with organic solvents, followed by drying [116]. These steps aim to eliminate cells, cellular debris, and residual nutrients, ensuring the purity and efficacy of the final product, especially when intended for sensitive applications in pharmaceutical, cosmetic, and food industries [113]. Structural characterization of purified compounds is also essential to elucidate their composition and functional properties. Techniques such as FTIR, LC-MS, and NMR are frequently used to confirm the presence of functional groups and molecular structure of biosurfactants [117, 118].

4. Mechanisms of action of biosurfactants in the mitigation of abiotic stress in plants

4.1 Ecological mechanisms

Biosurfactants play an important role in mitigating abiotic stress in plants through a comprehensive set of physical and chemical mechanisms that affect soil dynamics, the rhizosphere, and plant physiology (**Figure 2**). These compounds exert both direct and indirect effects on plants and their surrounding environment, conferring greater tolerance to adverse conditions such as salinity, drought, the presence of metals, organic contaminants, pH variations, and extreme temperatures [107]. The multifunctionality of biosurfactants makes them promising tools for sustainable agriculture, with their efficacy mainly anchored in complex yet highly effective physicochemical interactions [32, 119].

Regarding the physical properties of soil, biosurfactants promote significant changes that favor plant growth. By interacting with mineral particles and organic matter, they facilitate the formation of stable aggregates, increasing porosity and water retention capacity [120]. This structural reorganization improves aeration, reduces compaction, and stimulates root system development [121]. In environments with water deficit, for example, these compounds enhance water infiltration and redistribution in the soil, facilitating its uptake by roots and contributing to the maintenance of cell turgor and plant metabolism [70, 122].

Biosurfactants also positively influence the bioavailability of essential nutrients in soils, especially those degraded or with low fertility [123, 124]. Elements such as phosphorus, iron, manganese, potassium, and zinc, often retained in the soil matrix in forms that are poorly assimilable—particularly under adverse pH or salinity

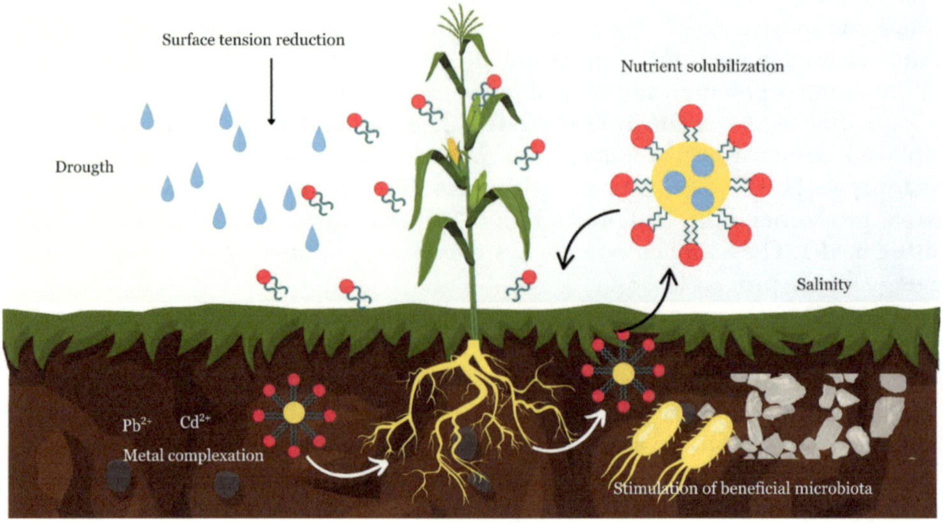

Figure 2.
Schematic illustrating the main physicochemical mechanisms by which biosurfactants contribute to the mitigation of abiotic stress in plants.

conditions—become more accessible to plants [73, 125]. The presence of negatively charged functional groups in biosurfactants contributes to increasing the soil's cation exchange capacity, thereby improving the retention and availability of these elements [34, 126]. Enhanced nutrient availability, combined with rhizosphere pH modulation, supports soil fertility and mineral nutrition in stressed environments [73, 125]. These effects underpin fundamental metabolic processes such as photosynthesis, respiration, and plant growth [127].

In the presence of organic or inorganic contaminants, biosurfactants act by forming micelles and emulsions that encapsulate hydrophobic molecules such as pesticides, hydrocarbons, and agro-industrial residues [7, 128], or by complexing metal ions, reducing their toxicity or facilitating their removal [129, 130]. This complexation and chelation ability occurs through interactions with functional groups such as carboxyls, phosphates, and hydroxyls present in the polar fraction of these molecules [73]. These groups serve as binding sites for cations such as Cd^{2+}, Pb^{2+}, Zn^{2+}, Cu^{2+}, and Ni^{2+} [53, 73, 131], preventing excessive accumulation of these elements in plant tissues, thereby preserving cellular integrity and enzymatic activity [132, 133].

Biosurfactants contribute to the stabilization of cellular membranes under osmotic, thermal, and oxidative stresses [71, 134], reducing lipid peroxidation and preventing solute leakage [134, 135]. This protection supports the proper functioning of ion pumps, cell signaling systems, and membrane transporters, all essential for maintaining osmotic and nutritional balance in plants [82, 128, 136].

The physicochemical properties of biosurfactants decisively contribute to plant resilience in adverse environments by reducing environmental toxicity, protecting cellular structures, and increasing efficiency in water and nutrient uptake. These compounds induce significant changes at the water-soil interface, influence rhizosphere microbial dynamics, and stabilize intracellular physiological processes, thereby creating favorable conditions for more complex biological and ecological mechanisms [137].

4.2 Biological mechanisms

Biosurfactants directly influence the rhizosphere microbiota the soil region surrounding and metabolically influenced by plant roots. Acting as ecological facilitators, these compounds promote root colonization by beneficial microorganisms, such as plant growth-promoting bacteria (PGPB) and mycorrhizal fungi [138]. The efficiency of the plant microbe interaction depends on the ability of bacteria to adhere to the rhizosphere and form structured communities such as biofilms. These biofilms protect microorganisms from adverse conditions such as drought and salinity, enhancing their survival and functional activity within the rhizosphere. Moreover, they act as microenvironments rich in biochemical exchanges, fostering cooperative interactions between microbial species and the host plant. This synergy results in a more resilient and functional root system under environmental constraints [10].

Biosurfactants play a crucial role in this process by promoting cell adhesion, bacterial motility, and the organization of the extracellular matrix that sustains the biofilm. Bacteria such as *Burkholderia* and *Pseudomonas* enhance both their own adhesion to root surfaces and the formation of biofilms, thereby strengthening the symbiotic relationship with the plant [139]. For the plant, this stable association ensures the continuous supply of beneficial metabolites, phytohormones, and enzymes—even during prolonged stress conditions [10].

Biosurfactants can act indirectly, by facilitating colonization by phytohormone-producing microorganisms, or directly, by modulating the expression of plant genes related to hormonal pathways [140]. Auxins, particularly indole-3-acetic acid (IAA), stimulate lateral and adventitious root growth. This root development increases the absorption surface area for water and nutrients, which is vital for plant survival under drought or saline stress. In addition to auxins, there is evidence that rhizospheric microorganisms also produce cytokinins, which delay leaf senescence, and gibberellins, which promote cell elongation [141, 142].

Microbial communities in the rhizosphere and biosurfactants also exert direct antimicrobial activity against phytopathogens, contributing to plant protection under stressful environments. Compounds such as lipopeptides (e.g., surfactin, fengycin, and iturin) and glycolipids (e.g., rhamnolipids) exhibit specific antimicrobial activity against phytopathogenic bacteria and fungi [139, 143]. These biosurfactants primarily act by permeabilizing the pathogen's cell membranes, disrupting the lipid bilayer and leading to cell lysis. Additionally, they interfere with the formation of pathogenic biofilms by inhibiting the adhesion of undesirable microorganisms to roots. This effect reduces colonization by species such as *Aspergillus sp.*, *Fusarium sp.*, *Pythium sp.*, and *Rhizoctonia solani*, while enhancing the competitiveness of beneficial microbes in the rhizoplane [139].

Thus, biosurfactants not only mediate biological interactions between plants, soil, and microorganisms but also trigger responses at the molecular level, including the modulation of signaling pathways, regulation of gene expression, and activation of physiological responses in plants under abiotic stress [140, 144].

4.3 Molecular mechanisms

Biosurfactants are capable of modulating molecular and physiological processes in plants, promoting adaptive responses that enhance resilience to abiotic stresses such as drought, salinity, extreme temperatures, metal toxicity, and nutrient deficiencies

[71, 107, 145]. Among the mechanisms mediated by these compounds are the induction of antioxidant gene expression, modulation of stress-associated hormonal pathways, and regulation of essential physiological functions, including photosynthesis, stomatal conductance, osmotic homeostasis, and ion balance [70, 146, 147].

At the molecular level, biosurfactants stimulate the transcription of genes associated with the activity of antioxidant enzymes such as superoxide dismutase (SOD), catalase (CAT), and peroxidases (POD), which are fundamental to maintaining redox balance and cellular integrity under abiotic stress [27, 148]. These enzymes eliminate reactive oxygen species (ROS), which, despite being potentially toxic, also serve as crucial secondary messengers in signaling cascades that activate physiological and adaptive responses. At controlled concentrations, ROS induce the expression of defense-related genes through the activation of specific transcription factors [108, 149].

The production and regulation of ROS are highly controlled processes. In plants, ROS are primarily generated by NADPH oxidase enzymes, known as respiratory burst oxidase homologs (RBOHs), mostly located on the plasma membrane but also present in organelles such as vacuoles, the endoplasmic reticulum, and mitochondria [146, 150]. These enzymes are regulated by post-translational modifications, such as phosphorylation mediated by calcium-dependent CBL-CIPK kinase complexes, enabling rapid and specific responses to environmental stress. Localized ROS production can also initiate redox signaling waves by activating RBOHs in neighboring cells, thereby propagating systemic defense signals [146, 151].

Among the ROS-regulated transcription factor families, WRKY, AP2/ERF, NAC, and bZIP stand out for their role in recognizing cis-regulatory elements in the promoter regions of antioxidant genes [108, 152, 153]. SOD converts superoxide radicals ($O_2^{-}\bullet$) into hydrogen peroxide (H_2O_2), which is subsequently degraded into H_2O and O_2 by the action of CAT and POD. This enzymatic coordination limits ROS accumulation and prevents oxidative damage to essential biomolecules such as DNA, lipids, and proteins [146, 154].

In this context, biosurfactants act as redox homeostasis modulators by inducing antioxidant responses in plants under stress. Evidence indicates that these compounds activate ROS-dependent signaling pathways, culminating in antioxidant gene induction and oxidative damage mitigation [27]. Additionally, biosurfactant application via soil or foliar routes can reduce proline levels—a classical marker of oxidative stress—by enhancing osmotic balance and activating antioxidant systems [148, 155].

Physiologically, plants treated with these compounds exhibit higher photosynthetic rates, stomatal conductance, and leaf water potential—key parameters for metabolic performance under adverse conditions. For example, Chen et al. [156] observed a significant increase in antioxidant capacity and protection of the photosynthetic apparatus in tea plants (*Camellia sinensis L.*) treated with biosurfactants under high-temperature stress.

Under saline stress, biosurfactants regulate ion transport by reducing sodium (Na^+) uptake and promoting potassium (K^+) accumulation—an essential element for enzymatic activity and cell turgor [113]. This modulation is partially due to the transcriptional regulation of ion channels and transporters. Studies in maize have shown that biosurfactants promote the biosynthesis of osmoprotectants (such as proline and glycine betaine) and increase the expression of aquaporins and specific transporters [12, 27].

At the hormonal level, microbial biosurfactants function as elicitors by modulating the biosynthesis and signaling pathways of key phytohormones involved in stress adaptation, including abscisic acid (ABA), salicylic acid (SA), and jasmonates (JA) [72]. These compounds interact with pattern recognition receptors (PRRs), triggering signaling cascades that elevate endogenous hormone levels [156, 157].

Abscisic acid, classically involved in drought and salinity responses, promotes stomatal closure and the transcription of protective genes such as *NCED3*, whose expression can be induced by biosurfactants [158]. Salicylic acid, associated with systemic acquired resistance (SAR), also plays a role in abiotic stress responses and can induce genes like *PR1* [150, 159]. Regulation of this pathway involves transcriptional complexes with NPR1, TGA, and histone acetyltransferases (HACs), which promote epigenetic reprogramming via histone acetylation—crucial for gene activation [160]. Jasmonates, in turn, are essential for defense against necrotrophic pathogens and herbivory, with their signaling being amplified by biosurfactants through the activation of genes such as *LOX*, *AOS*, and *JAZ*. JA pathways can interact synergistically or antagonistically with other signaling routes, depending on the type and intensity of the stressor [161].

In an integrated manner, biosurfactant action on antioxidant, hormonal, and ion transport systems strengthens induced systemic resistance (ISR), promoting a more efficient multiscale response to environmental adversity [156]. Furthermore, functional genomics of biosurfactant-producing microorganisms has enabled the identification of key genes involved in the synthesis of biosurfactants with high specificity and agricultural efficacy [162, 163].

Mechanistic understanding through omics technologies—such as transcriptomics, proteomics, metabolomics, and genomics—has been essential to elucidating the molecular mechanisms underlying microbial biosurfactant action in plants under abiotic stress. These approaches allow the integrated identification of physiological and biochemical responses triggered after compound application and support the selection of more effective strains and formulations for specific agricultural contexts [164, 165]. Metabolomics techniques such as MS and NMR have revealed the accumulation of osmoprotectants like proline and trehalose in biosurfactant-treated plants, particularly under salinity and drought [78]. Proteomics, in turn, has shown changes in proteins related to photosynthesis and defense, while functional genomics has identified genes associated with the production of biosurfactants with greater agricultural efficacy [166]. These data, integrated with phenotypic assays, have demonstrated direct correlations between gene expression and plant agronomic performance, strengthening the use of biosurfactants as bio-stimulants tailored to diverse crops and environmental conditions.

The complexity and synergy of these mechanisms highlight the enormous potential of these molecules for sustainable agricultural applications, especially in contexts of environmental degradation, water scarcity, or contamination [167]. Further research into biosurfactant structure, optimal concentration, interactions with different plant species, and application methods may further consolidate their role as promising tools for enhancing plant resilience and productivity in modern agroecological systems.

5. Technological challenges and advances in the production and application of microbial biosurfactants under abiotic stress conditions

Despite technological progress, the commercial-scale production of biosurfactants still faces major barriers, mainly low yields, expensive substrates, complex purification steps, and high costs—up to 30% higher than chemical surfactants [168, 169]. Variability in physicochemical properties between strains and fermentation conditions also complicates standardization and reproducibility [170].

Advances in industrial biotechnology have helped address these challenges through innovations in fermentation (continuous systems, membrane aeration, and statistical optimization) and in downstream processing (selective membranes and sustainable precipitation methods), improving yield and quality [5, 84, 108, 118, 171]. The use of low-cost substrates such as agro-industrial residues is essential for reducing expenses, adding competitiveness, and valorizing waste [107, 109].

Since many microbial species are unculturable, high-throughput screening and metagenomics have expanded bioprospecting, enabling the discovery of novel biosurfactant-producing microorganisms and compounds with unique properties [84, 150]. Genetic engineering further enhances production by optimizing strains for yield, stability, and tolerance to stress. Tools such as recombinant DNA technology and CRISPR-Cas9 have been used to overexpress biosynthetic genes, inhibit competing pathways, and improve cellular robustness [172–174]. Genetic modifications in model strains like *Pseudomonas*, *Bacillus*, and *Escherichia* already demonstrate significant productivity gains [175], while functional genomics supports the selection of strains for agricultural applications.

Beyond scientific advances, external drivers also influence the field. The demand for sustainable agricultural inputs and environmentally responsible practices has fostered green technologies and shaped public policies in the bioeconomy [34, 114, 176]. Incentives, funding, and academia-industry partnerships have accelerated the technological maturity of biosurfactants [177–179].

Ultimately, the economic feasibility of biosurfactants will determine their adoption. Consolidating cost-effective and environmentally sustainable production platforms, aligned with science, innovation, and policy, will be crucial for their integration into agriculture, waste valorization, and the transition to a greener economy [34, 84, 114, 176].

6. Conclusion

The global market for microbial surfactants has grown significantly in recent decades due to the excellent functional properties of these biomolecules, which enable their application in bioremediation, medicine, cosmetics, and food industries, among others. In the face of increasing challenges posed by abiotic stressors in agriculture, microbial biosurfactants have emerged as promising tools to enhance plant resilience and restore degraded ecosystems. Through physical, chemical, biological, and molecular mechanisms, these biomolecules facilitate nutrient uptake, protect plants against adverse conditions, and positively modulate rhizospheric microbiota. Although certain obstacles still limit their large-scale application, advances in biotechnology and sustainable production point to a promising future in which biosurfactants will play a central role in building a more sustainable agriculture one that is better adapted to climate change and less reliant on synthetic inputs.

Author details

Bruna Sthefanie Paz de Souza[1], Glacielen Ribeiro de Souza[1], Analyse Villanueva Gaete[1], Gabriela Petroceli Mota[2], João Vitor Pereira de Souza[2], Daiane Carvalho Baía[1], Marília Berbert Molina[1] and Aline Chaves Intorne[1,3*]

1 Biotechnology Laboratory (LBT), Universidade Estadual do Norte Fluminense Darcy Ribeiro (UENF), Rio de Janeiro, Brasil

2 Cellular and Tissue Biology Laboratory (LBCT), Universidade Estadual do Norte Fluminense Darcy Ribeiro (UENF), Rio de Janeiro, Brasil

3 Federal Institute of Education Science and Technology of Rio de Janeiro, Rio de Janeiro, Brasil

*Address all correspondence to: aline.intorne@ifrj.edu.br

References

[1] Elias C, Ranque S, Malleret L. State of the art on fungal biodegradation of persistent organic pollutants in soils and innovative strategies for isolating relevant candidate strains. Environmental Technology & Innovation. 2025;**39**:104247. DOI: 10.1016/j.eti.2025.104247

[2] Srivastava RK, Bothra N, Singh R, Sai MC, Nedungadi SV, Sarangi PK. Microbial originated surfactants with multiple applications: A comprehensive review. Archives of Microbiology. 2022;**204**(8):452. DOI: 10.1007/s00203-022-03086-3

[3] Ribeiro C, Hennen-Bierwagen TA, Myers AM, Cline K, Settles AM. Engineering 6-phosphogluconate dehydrogenase improves grain yield in heat-stressed maize. Proceedings of the National Academy of Sciences. 2020;**117**(52):33177-33185. DOI: 10.1073/pnas.2010179117

[4] Rzymski P, Gwenzi W, Poniedziałek B, Mangul S, Fal A. Climate warming, environmental degradation and pollution as drivers of antibiotic resistance. Environmental Pollution. 2024;**346**:123649. DOI: 10.1016/j.envpol.2024.123649

[5] Narayanan M, Ma Y. Mitigation of heavy metal stress in the soil through optimized interaction between plants and microbes. Journal of Environmental Management. 2023;**345**:118732. DOI: 10.1016/j.jenvman.2023.118732

[6] Marchant R, Banat IM. Biosurfactants: A sustainable replacement for chemical surfactants? Biotechnology Letters. 2012;**34**:1597-1605. DOI: 10.1007/s10529-012-0956-x

[7] Hassan MA, Xiang C, Farooq M, Muhammad N, Yan Z, Hui X, et al. Cold stress in wheat: Plant acclimation responses and management strategies. Frontiers in Plant Science. 2021;**12**. DOI: 10.3389/fpls.2021.676884

[8] Rivero M, Gutierrez-Cacciabue D, Sanguino-Jorquera DG, Rajal VB, Irazusta VP. High throughput screening of the potential biosurfactants production by extremophiles isolated from vinasse and black liquor. Results in Engineering. 2022;**15**:100587. DOI: 10.1016/j.rineng.2022.100587

[9] Daâssi D, Almaghrabi FQ. Petroleum-degrading fungal isolates for the treatment of soil microcosms. Microorganisms. 2023;**11**(5):1351. DOI: 10.3390/microorganisms11051351

[10] Castelein M, de Fooz N, Luyten G, Van Renterghem L, Dierickx S, Bovijn S, et al. Improved production of novel (bola) glycolipid biosurfactants with the yeast *Starmerella bombicola* through an integrative approach combining genetic engineering and multiomics analyses. In: Biosurfactants. Amsterdam, Netherlands: Elsevier; 2023. pp. 183-202. DOI: 10.1016/B978-0-323-91697-4.00009-0

[11] Kopecká R, Kameniarová M, Černý M, Brzobohatý B, Novák J. Abiotic stress in crop production. International Journal of Molecular Sciences. 2023;**24**(7):6603. DOI: 10.3390/ijms24076603

[12] Gutierrez MM, Cameron-Harp MV, Chakraborty PP, Stallbaumer-Cyr EM, Morrow JA, Hansen RR, et al. Investigating a microbial approach to water conservation: Effects of *Bacillus subtilis* and Surfactin on evaporation

dynamics in loam and sandy loam soils. Frontiers in Sustainable Food Systems. 2022;**6**:959591. DOI: 10.3389/fsufs.2022.959591

[13] do Vale KV, Soares DWF, da Silva ALB, da Abreu FOMS, de Oliveira Almeida JLI, Oliveira MRF, et al. Biossurfactant production by isolated bacteria from petroleum industry's effluent targeting bioremediation. Results in Chemistry. 2023;**6**:101038. DOI: 10.1016/j.rechem.2023.101038

[14] Dasgan HY, Dere S, Akhoundnejad Y, Arpaci BB. Effects of high-temperature stress during plant cultivation on tomato (*Solanum lycopersicum* L.) fruit nutrient content. Journal of Food Quality. 2021;**2021**:1-15. DOI: 10.1155/2021/7994417

[15] Onwosi CO, Odibo FJC. Effects of carbon and nitrogen sources on rhamnolipid biosurfactant production by pseudomonas nitroreducens isolated from soil. World Journal of Microbiology and Biotechnology. 2012;**28**:937-942. DOI: 10.1007/s11274-011-0891-3

[16] Osman MS, Ibrahim Z, Japper-Jaafar A, Shahir S. Biossurfactants and its prospective application in the petroleum industry. Journal of Sustainability Science and Management. 2019;**14**:125-140. Available from: https://api.semanticscholar.org/CorpusID:231582237

[17] Ferreira MJ, Veríssimo ACS, Pinto DCGA, Sierra-Garcia IN, Granada CE, Cremades J, et al. Engineering the rhizosphere microbiome with plant growth promoting bacteria for modulation of the plant metabolome. Plants. 2024;**13**(16):2309. DOI: 10.3390/plants13162309

[18] Silva JHCS, Rodrigues CM, das Souza A, do Nascimento GNFF, Alves EU. Effects of temperature and salt stress on *Cereus fernambucensis* seed germination. Biology. 2025;**14**(4):393. DOI: 10.3390/biology14040393

[19] Hosseini H, Al-Jabri HM, Moheimani NR, Siddiqui SA, Saadaoui I. Marine microbial bioprospecting: Exploitation of marine biodiversity towards biotechnological applications—A review. Journal of Basic Microbiology. 2022;**62**:1030-1043. DOI: 10.1002/jobm.202100504

[20] Khan BA, Nadeem MA, Nawaz H, Amin MM, Abbasi GH, Nadeem M, et al. Pesticides: Impacts on agriculture productivity, environment, and management strategies. In: Emerging Contaminants and Plants: Interactions, Adaptations and Remediation Technologies. Cham, Switzerland: Springer International Publishing; 2023. pp. 109-134. DOI: 10.1007/978-3-031-22269-6_5

[21] Luna JM, Rufino RD, Jara AMAT, Brasileiro PPF, Sarubbo LA. Environmental applications of the biosurfactant produced by *Candida sphaerica* cultivated in low-cost substrates. Colloids and Surfaces A: Physicochemical and Engineering Aspects. 2015;**480**:413-418. DOI: 10.1016/j.colsurfa.2014.12.014

[22] Shah V, Dani P, Daverey A. Phytoremediation of heavy metal contaminated soil using Bidens pilosa: Effect of varying concentrations of Sophorolipids. Applied Biochemistry and Biotechnology. 2024;**196**(5):2399-2413. DOI: 10.1007/s12010-023-04684-7

[23] Min X, Zhiyong X, Zhuang X, Zhenya W. Analysis of enablers within environmental policy frameworks for facilitating circular economy practices in the sports products industry. Journal of Environmental Management.

2024;**372**:123306. DOI: 10.1016/j.jenvman.2024.123306

[24] Naeem M, Gill R, Gill SS, Singh K, Sofo A, Tuteja N. Editorial: Emerging contaminants and their effect on agricultural crops. Frontiers in Plant Science. 2023;**14**. DOI: 10.3389/fpls.2023.1296252

[25] Rajametov SN, Lee K, Jeong H-B, Cho M-C, Nam C-W, Yang E-Y. Physiological traits of thirty-five tomato accessions in response to low temperature. Agriculture. 2021;**11**(8):792. DOI: 10.3390/agriculture11080792

[26] Shankar T, Praharaj S, Sahoo U, Maitra S. Intensive farming: Its effect on the environment. Indian Journal of Natural Sciences. 2021;**12**:37480-37487

[27] Ali B, Wang X, Saleem MH, Sumaira, Hafeez A, Afridi MS, et al. PGPR-mediated salt tolerance in maize by modulating plant physiology, antioxidant defense, compatible solutes accumulation and bio-surfactant producing genes. Plants. 2022;**11**(3):345. DOI: 10.3390/plants11030345

[28] Thakur P, Saini NK, Thakur VK, Gupta VK, Saini RV, Saini AK. Rhamnolipid the glycolipid biosurfactant: Emerging trends and promising strategies in the field of biotechnology and biomedicine. Microbial Cell Factories. 2021;**20**(1):1. DOI: 10.1186/s12934-020-01497-9

[29] Tison L, Beaumelle L, Monceau K, Thiéry D. Transfer and bioaccumulation of pesticides in terrestrial arthropods and food webs: State of knowledge and perspectives for research. Chemosphere. 2024;**357**:142036. DOI: 10.1016/j.chemosphere.2024.142036

[30] Imran QM, Falak N, Hussain A, Mun B-G, Yun B-W. Abiotic stress in plants; stress perception to molecular response and role of biotechnological tools in stress resistance. Agronomy. 2021;**11**(8):1579. DOI: 10.3390/agronomy11081579

[31] IPCC. Climate change 2023: Synthesis report. In: Core Writing Team, Lee H, Romero J, editors. Contribution of Working Groups I, II and III to the Sixth Assessment Report of the Intergovernmental Panel on Climate Change. Geneva, Switzerland: IPCC; 2023

[32] Mahmoud YA-G, El-Halmouch YH, Nasr EE, Al-Sodany YM, El-Nogoumy BA, Ali SS. Exploring the potential of biosurfactants produced by fungi found in soil contaminated with petrochemical wastes. Scientific Reports. 2024;**14**(1):25733. DOI: 10.1038/s41598-024-75865-5

[33] de Soares da Silva RCF, Luna JM, Rufino RD, Sarubbo LA. Ecotoxicity of the formulated biosurfactant from *Pseudomonas cepacia* CCT 6659 and application in the bioremediation of terrestrial and aquatic environments impacted by oil spills. Process Safety and Environmental Protection. 2021;**154**:338-347. DOI: 10.1016/j.psep.2021.08.038

[34] Mathur S, Agnihotri R, Sharma MP, Reddy VR, Jajoo A. Effect of high-temperature stress on plant physiological traits and Mycorrhizal Symbiosis in maize plants. Journal of Fungi. 2021;7(10):867. DOI: 10.3390/jof7100867

[35] Zhu T, Fonseca De Lima CF, De Smet I. The heat is on: How crop growth, development, and yield respond to high temperature. Journal of Experimental Botany. 20 Nov 2021;**72**(21):7359-7373. DOI: 10.1093/jxb/erab308

[36] Chamkhi I, El Omari N, Balahbib A, El Menyiy N, Benali T,

Ghoulam C. Is —— The rhizosphere a source of applicable multi-beneficial microorganisms for plant enhancement? Saudi Journal of Biological Sciences. 2022;**29**(2):1246-1259. DOI: 10.1016/j.sjbs.2021.09.032

[37] Zhao Y, Wang J, Huang W, Zhang D, Wu J, Li B, et al. Abscisic-acid-regulated responses to alleviate cadmium toxicity in plants. Plants. 2023;**12**(5):1023. DOI: 10.3390/plants12051023

[38] Wang Y, Wan S, Yu W, Yuan D, Sun L. Newly isolated *Enterobacter cloacae* sp. HN01 and *Klebsiella pneumoniae* sp. HN02 collaborate with self-secreted biosurfactant to improve solubility and bioavailability for the biodegradation of hydrophobic and toxic gaseous Para-xylene. Chemosphere. 2022;**304**:135328. DOI: 10.1016/j.chemosphere.2022.135328

[39] Dini S, Bekhit AE-DA, Roohinejad S, Vale JM, Agyei D. The physicochemical and functional properties of biosurfactants: A review. Molecules. 2024;**29**(11):2544. DOI: 10.3390/molecules29112544

[40] Jin H, Choi S-M, Kang M-J, Yun S-H, Kwon D-J, Noh Y-S. Salicylic acid-induced transcriptional reprogramming by the HAC–NPR1–TGA histone acetyltransferase complex in Arabidopsis. Nucleic Acids Research. 2018. DOI: 10.1093/nar/gky847

[41] Habib S, Ahmed A, Haque MFU, Ejaz R. Efficacious use of potential biosurfactant producing plant growth promoting rhizobacteria to combat petrol toxicity in *Zea mays* L. plants. Environmental Science and Pollution Research. 2023;**30**(18):53725-53740. DOI: 10.1007/s11356-023-25925-6

[42] Gupta S, Kaur R, Sharma T, Bhardwaj A, Sharma S, Sohal JS, et al. Multi-omics approaches for understanding stressor-induced physiological changes in plants: An updated overview. Physiological and Molecular Plant Pathology. 2023;**126**:102047. DOI: 10.1016/j.pmpp.2023.102047

[43] Kosová K, Nešporová T, Vítámvás P, Vítámvás J, Klíma M, Ovesná J, et al. How to survive mild winters: Cold acclimation, deacclimation, and reacclimation in winter wheat and barley. Plant Physiology and Biochemistry. 2025;**220**:109541. DOI: 10.1016/j.plaphy.2025.109541

[44] Put H, Gerstmans H, Vande Capelle H, Fauvart M, Michiels J, Masschelein J. Bacillus subtilis as a host for natural product discovery and engineering of biosynthetic gene clusters. Natural Product Reports. 2024;**41**(7):1113-1151. DOI: 10.1039/D3NP00065F

[45] Tancredi M, Carandente Coscia C, Russo Krauss I, D'Errico G. Antioxidant properties of biosurfactants: Multifunctional biomolecules with added value in formulation chemistry. Biomolecules. 2025;**15**(2):308. DOI: 10.3390/biom15020308

[46] Zhou X, Muhammad I, Lan H, Xia C. Recent advances in the analysis of cold tolerance in maize. Frontiers in Plant Science. 2022;**13**. DOI: 10.3389/fpls.2022.866034

[47] Kohrt JR, Sprague CL, Swathi Nadakuduti S, Douches D. Confirmation of a three-way (glyphosate, ALS, and atrazine) herbicide-resistant population of palmer Amaranth (*Amaranthus palmeri*) in Michigan. Weed Science. 2017;**65**(3):327-338. DOI: 10.1017/wsc.2017.2

[48] Kumar A, Singh SK, Kant C, Verma H, Kumar D, Singh PP, et

al. Microbial biosurfactant: A new Frontier for sustainable agriculture and pharmaceutical industries. Antioxidants. 2021;**10**(9):1472. DOI: 10.3390/antiox10091472

[49] Waheed A, Xu H, Qiao X, Aili A, Yiremaikebayi Y, Haitao D, et al. Biochar in sustainable agriculture and climate mitigation: Mechanisms, challenges, and applications in the circular bioeconomy. Biomass and Bioenergy. 2025;**193**:107531. DOI: 10.1016/j.biombioe.2024.107531

[50] Akbari E, Rasekh B, Maal KB, Karbasiun F, Yazdian F, Emami-Karvani Z, et al. A novel biosurfactant producing Kocuria rosea ABR6 as potential strain in oil sludge recovery and lubrication. AMB Express. 2021;**11**:61. DOI: 10.1186/s13568-021-01283-9

[51] Ghori N-H, Ghori T, Hayat MQ, Imadi SR, Gul A, Altay V, et al. Heavy metal stress and responses in plants. International journal of Environmental Science and Technology. 2019;**16**(3):1807-1828. DOI: 10.1007/s13762-019-02215-8

[52] Almansoory AF, Hasan HA, Idris M, Abdullah SRS, Anuar N, Tibin EMM. Biosurfactant production by the hydrocarbon-degrading bacteria (HDB) *Serratia marcescens*: Optimization using central composite design (CCD). Journal of Industrial and Engineering Chemistry. 2017;**47**:272-280. DOI: 10.1016/j.jiec.2016.11.005

[53] Shawky E, Gibbons S, Selim DA. Bio-sourcing from byproducts: A comprehensive review of bioactive molecules in Agri-food waste (AFW) streams for valorization and sustainable applications. Bioresource Technology. 2025;**431**:132640. DOI: 10.1016/j.biortech.2025.132640

[54] de Souza GR, Petroceli-Mota G, Gaete AV, de Souza BSP, Berbert-Molina MA, Intorne AC. Synthetic microbial community as an innovative strategy for bioremediation of industrial effluents. In: Advances in Applied Microbiology. Amsterdam, Netherlands: Elsevier; 2025. pp. 391-418. DOI: 10.1016/bs.apmp.2024.10.009

[55] Dietz K-J, Zörb C, Geilfus C-M. Drought and crop yield. Plant Biology. 2021;**23**(6):881-893. DOI: 10.1111/plb.13304

[56] Gebregiorgis Ambaye T, Vaccari M, Franzetti A, Prasad S, Formicola F, Rosatelli A, et al. Microbial electrochemical bioremediation of petroleum hydrocarbons (PHCs) pollution: Recent advances and outlook. Chemical Engineering Journal. 2023;**452**:139372. DOI: 10.1016/j.cej.2022.139372

[57] Rashid A, Schutte BJ, Ulery A, Deyholos MK, Sanogo S, Lehnhoff EA, et al. Heavy metal contamination in agricultural soil: Environmental pollutants affecting crop health. Agronomy. 2023;**13**(6):1521. DOI: 10.3390/agronomy13061521

[58] Fardin AB, Jamshidi-Zanjani A. A critical review on soil remediation using electrokinetic-enhanced permeable reactive barriers: Challenges and enhancements. Chemical Engineering Journal Advances. 2025;**23**:100774. DOI: 10.1016/j.ceja.2025.100774

[59] Zhang M, Duan T, Luo Y, Zhang H, Li W, Wang X, et al. Impact mechanisms of various surfactants on the biodegradation of phenanthrene in soil: Bioavailability and microbial community responses. Science of the Total Environment. 2024;**950**:175225. DOI: 10.1016/j.scitotenv.2024.175225

[60] Thakur V, Baghmare P, Verma A, Verma JS, Geed SR. Recent

progress in microbial biosurfactants production strategies: Applications, technological bottlenecks, and future outlook. Bioresource Technology. 2024;**408**:131211. DOI: 10.1016/j.biortech.2024.131211

[61] Sharma A, Kumar V, Thukral AK, Bhardwaj R. Responses of plants to pesticide toxicity: An overview. Planta Daninha. 2019;**37**:65. DOI: 10.1590/s0100-83582019370100065

[62] Kashif A, Rehman R, Fuwad A, Shahid MK, Dayarathne N, Jamal A, et al. Current advances in the classification, production, properties and applications of microbial biosurfactants – A critical review. Advances in Colloid and Interface Science. 2022;**306**:102718

[63] Worakittikul W, Srisathan WA, Rattanpon K, Kulkaew A, Groves J, Pontun P, et al. Cultivating sustainability: Harnessing open innovation and circular economy practices for eco-innovation in agricultural SMEs. Journal of Open Innovation: Technology, Market, and Complexity. 2025;**11**(1):100494. DOI: 10.1016/j.joitmc.2025.100494

[64] Tripathi L, Twigg MS, Zompra A, Salek K, Irorere VU, Gutierrez T, et al. Biosynthesis of rhamnolipid by a *Marinobacter* species expands the paradigm of biosurfactant synthesis to a new genus of the marine microflora. Microbial Cell Factories. 2019;**18**:1-12. DOI: 10.1186/s12934-019-1216-8

[65] Banat IM, Carboué Q, Saucedo-Castañeda G, de Jesús Cázares-Marinero J. Biosurfactants: The green generation of speciality chemicals and potential production using solid-state fermentation (SSF) technology. Bioresource Technology. 2021;**320**:124222. DOI: 10.1016/j.biortech.2020.124222

[66] El-Wakeil N, Gaafar N, Sallam A, Volkmar C. Side effects of insecticides on natural enemies and possibility of their integration in plant protection strategies. In: Insecticides - Development of Safer and more Effective Technologies. London, UK: IntechOpen; 2013. DOI: 10.5772/54199

[67] Mittler R. ROS are good. Trends in Plant Science. 2017;**22**(1):11-19. DOI: 10.1016/j.tplants.2016.08.002

[68] Ketemepi HK, Awang MA, Bin S, Seelan JSS, Mohd Noor NQI. Extraction process and applications of mushroom-derived protein hydrolysate: A comprehensive review. Future Foods. 2024;**9**:100359. DOI: 10.1016/j.fufo.2024.100359

[69] Perera HCS, Gurunanthanan V, Singh A, Mantilaka MMMGPG, Das G, Arya S. Magnesium oxide (MgO) nanoadsorbents in wastewater treatment: A comprehensive review. Journal of Magnesium and Alloys. 2024;**12**(5):1709-1773. DOI: 10.1016/j.jma.2024.05.003

[70] Abhyankar I, Hirlekar S, Prabhune A, Nisal A. Bridging the gap: An investigation of biosurfactants-polymer systems. Current Opinion in Colloid & Interface Science. 2024;**72**:101806. DOI: 10.1016/j.cocis.2024.101806

[71] Eras-Muñoz E, Farré A, Sánchez A, Font X, Gea T. Microbial biosurfactants: A review of recent environmental applications. Bioengineered. 2022;**13**(5):12365-12391. DOI: 10.1080/21655979.2022.2074621

[72] Kugaji M, Ray SK, Parvatikar P, Raghu AV. Biosurfactants: A review of different strategies for economical production, their applications and recent advancements. Advances in Colloid and Interface Science. 2025;**337**:103389. DOI: 10.1016/j.cis.2024.103389

[73] Etesami H, Jeong BR, Glick BR. Potential use of *Bacillus* spp. as an effective biostimulant against abiotic stresses in crops—A review. Current Research in Biotechnology. 2023;**5**:100128. DOI: 10.1016/j.crbiot.2023.100128

[74] Baek D, Rokibuzzaman M, Khan A, Kim MC, Park HJ, Yun D, et al. Plant-growth promoting *Bacillus oryzicola* YC7007 modulates stress-response gene expression and provides protection from salt stress. Frontiers in Plant Science. 2020;**10**:1646. DOI: 10.3389/fpls.2019.01646

[75] Song Z, Liu J, Han J. Chromatin remodeling factors regulate environmental stress responses in plants. Journal of Integrative Plant Biology. 2021;**63**(3):438-450. DOI: 10.1111/jipb.13064

[76] Faccioli YE da S, de Oliveira KW, Campos-Guerra JM, Converti A, Soares da Silva R de CF, Sarubbo LA. Biosurfactants: Chemical properties, ecofriendly environmental applications, and uses in the industrial energy sector. Energies. 2024;**17**(20):5042. DOI: 10.3390/en17205042

[77] da Silva AF, Banat IM, Giachini AJ, Robl D. Fungal biosurfactants, from nature to biotechnological product: Bioprospection, production and potential applications. Bioprocess and Biosystems Engineering. 2021;**44**(10):2003-2034. DOI: 10.1007/s00449-021-02597-5

[78] Bhadra S, Chettri D, Kumar Verma A. Biosurfactants: Secondary metabolites involved in the process of bioremediation and biofilm removal. Applied Biochemistry and Biotechnology. 2023;**195**(9):5541-5567. DOI: 10.1007/s12010-022-03951-3

[79] Uddin M, Swathi KV, Anil A, Boopathy R, Ramani K, Sekaran G. Biosequestration of lignin in municipal landfill leachate by tailored cationic lipoprotein biosurfactant through *Bacillus tropicus* valorized tannery solid waste. Journal of Environmental Management. 2021;**300**:113755. DOI: 10.1016/j.jenvman.2021.113755

[80] Durval IJB, Mendonça AHR, Rocha IV, Luna JM, Rufino RD, Converti A, et al. Production, characterization, evaluation and toxicity assessment of a *Bacillus cereus* UCP 1615 biosurfactant for marine oil spills bioremediation. Marine Pollution Bulletin. 2020;**157**:111357. DOI: 10.1016/j.marpolbul.2020.111357

[81] Datta M, Chattopadhyay I. Applications of microbial biosurfactants in human health and environmental sustainability: A narrative review. Discover Medicine. 2024;**1**(1):160. DOI: 10.1007/s44337-024-00181-w

[82] de Araujo LLGC, Sodré LGP, Brasil LR, Domingos DF, de Oliveira VM, da Cruz GF. Microbial enhanced oil recovery using a biosurfactant produced by *Bacillus safensis* isolated from mangrove microbiota - Part I biosurfactant characterization and oil displacement test. Journal of Petroleum Science and Engineering. 2019;**180**:950-957. DOI: 10.1016/j.petrol.2019.06.031

[83] Alengebawy A, Abdelkhalek ST, Qureshi SR, Wang M-Q. Heavy metals and pesticides toxicity in agricultural soil and plants: Ecological risks and human health implications. Toxics. 2021;**9**(3):42. DOI: 10.3390/toxics9030042

[84] Thakur P, Kumar S, Malik JA, Berger JD, Nayyar H. Cold stress effects on reproductive development in grain crops: An overview. Environmental and Experimental Botany. 2010;**67**(3):429-443. DOI: 10.1016/j.envexpbot.2009.09.004

[85] Argentin MN, Martins LF, Sousa MP, Bossolan NRS. Biosurfactant from a thermo-halophilic strain of *Bacillus alveayuensis* isolated from a Brazilian oil reservoir: Production, chemical characterization, antimicrobial activity, and efficiency in wettability reversal and oil removal from oil-soaked sand. Geoenergy Science and Engineering. 2023;**231**:212324. DOI: 10.1016/j.geoen.2023.212324

[86] Sadok W, Jagadish SVK. The hidden costs of Nighttime warming on yields. Trends in Plant Science. 2020;**25**(7):644-651. DOI: 10.1016/j.tplants.2020.02.003

[87] da Silva RR, Santos JCV, Meira HM, Almeida SM, Sarubbo LA, Luna JM. Microbial biosurfactant: *Candida bombicola* as a potential remediator of environments contaminated by heavy metals. Microorganisms. 2023;**11**(11):2772. DOI: 10.3390/microorganisms11112772

[88] Luo J, Yang Z, Zhang F, Li C. Effect of nitrogen application on enhancing high-temperature stress tolerance of tomato plants during the flowering and fruiting stage. Frontiers in Plant Science. 2023;**14**:1172078. DOI: 10.3389/fpls.2023.1172078

[89] Frempong KEB, Wang J, Chen Y, Liu X, Frempong ANA, Liang X, et al. Amphipathic regulation of cottonseed protein with the conjugation of bayberry tannin for efficient surface decontamination of uranium. Journal of Hazardous Materials. 2025;**488**:137450. DOI: 10.1016/j.jhazmat.2025.137450

[90] Karim A, Osse EF, Khalloufi S. Innovative strategies for valorization of byproducts from soybean industry: A review on status, challenges, and sustainable approaches towards zero-waste processing systems. Heliyon. 2025;**11**(3):e42118. DOI: 10.1016/j.heliyon.2025.e42118

[91] Sundaram T, Govindarajan RK, Vinayagam S, Krishnan V, Nagarajan S, Gnanasekaran GR, et al. Advancements in biosurfactant production using agro-industrial waste for industrial and environmental applications. Frontiers in Microbiology. 2024;**15**:1357302. DOI: 10.3389/fmicb.2024.1357302

[92] Shahid MK, Batool A, Kashif A, Nawaz MH, Aslam M, Iqbal N, et al. Biofuels and biorefineries: Development, application and future perspectives emphasizing the environmental and economic aspects. Journal of Environmental Management. 2021;**297**:113268. DOI: 10.1016/j.jenvman.2021.113268

[93] Qian R, Li Y, Liu Y, Sun N, Liu L, Lin X, et al. Integrated transcriptomic and metabolomic analysis reveals the potential mechanisms underlying indium-induced inhibition of root elongation in wheat plants. Science of the Total Environment. 2024;**908**:168477. DOI: 10.1016/j.scitotenv.2023.168477

[94] Tudi M, Daniel Ruan H, Wang L, Lyu J, Sadler R, Connell D, et al. Agriculture development, pesticide application and its impact on the environment. International Journal of Environmental Research and Public Health. 2021;**18**(3):1112. DOI: 10.3390/ijerph18031112

[95] Nascimento MF, Keković P, Ribeiro IAC, Faria NT, Ferreira FC. Novel organic solvent nanofiltration approaches for microbial biosurfactants downstream processing. Membranes. 2023;**13**(1):81. DOI: 10.3390/membranes13010081

[96] Raza A, Mubarik MS, Sharif R, Habib M, Jabeen W, Zhang C, et al. Developing drought-smart, ready-to-grow future crops. The Plant Genome. 2023;**16**(1):20279. DOI: 10.1002/tpg2.20279

[97] Santos DKF, Rufino RD, Luna JM, Santos VA, Sarubbo LA. Biosurfactants: Multifunctional biomolecules of the 21st century. International Journal of Molecular Sciences. 2016;**17**:401. DOI: 10.3390/ijms17030401

[98] Arslan NP, Azad F, Orak T, Budak-Savas A, Ortucu S, Dawar P, et al. A review on bacteria-derived antioxidant metabolites: Their production, purification, characterization, potential applications, and limitations. Archives of Pharmacal Research. 2025;**48**(4):253-292. DOI: 10.1007/s12272-025-01541-5

[99] Ahmed M, Tóth Z, Decsi K. The impact of salinity on crop yields and the confrontational behavior of transcriptional regulators, nanoparticles, and antioxidant defensive mechanisms under stressful conditions: A review. International Journal of Molecular Sciences. 2024;**25**(5):2654. DOI: 10.3390/ijms25052654

[100] Bawa G, Feng L, Yan L, Du Y, Shang J, Sun X, et al. Pre-treatment of salicylic acid enhances resistance of soybean seedlings to *Fusarium solani*. Plant Molecular Biology. 2019;**101**(3):315-323. DOI: 10.1007/s11103-019-00906-x

[101] Karmakar R, Aggarwal S, Kathuria D, Singh N, Tripathi V, Sharma PK, et al. Valorization of food waste stream by harnessing bioactive compounds: A comprehensive review on the process, challenges and solutions. Food Bioscience. 2025;**69**:106833. DOI: 10.1016/j.fbio.2025.106833

[102] Ma Y, Freitas H, Dias MC. Strategies and prospects for biostimulants to alleviate abiotic stress in plants. Frontiers in Plant Science. 2022;**13**:1024243. DOI: 10.3389/fpls.2022.1024243

[103] Jamil M, Malook I, Rehman SU, Aslam MM, Fayyaz M, Shah G, et al. Inoculation of heavy metal resistant bacteria alleviated heavy metal-induced oxidative stress biomarkers in spinach (*Spinacia oleracea* L.). BMC Plant Biology. 2024;**24**(1):221. DOI: 10.1186/s12870-024-04757-7

[104] Nercessian D, Di Meglio L, De Castro R, Paggi R. Exploring the multiple biotechnological potential of halophilic microorganisms isolated from two Argentinean salterns. Extremophiles. 2015;**19**:1133-1143. DOI: 10.1007/s00792-015-0785-7

[105] Irandoost E, Ghorbanpour A, Jamkhaneh HB, Ghasemi P, Tirkolaee EB. Sustainable-circular water resources management in the mining industry using system dynamics modeling. Water Resources and Industry. 2025;**34**:100313. DOI: 10.1016/j.wri.2025.100313

[106] Gayathiri E, Prakash P, Karmegam N, Varjani S, Awasthi MK, Ravindran B. Biosurfactants: Potential and eco-friendly material for sustainable agriculture and environmental safety—A review. Agronomy. 2022;**12**(3):662. DOI: 10.3390/agronomy12030662

[107] Garay LA, Sitepu IR, Cajka T, Xu J, Teh HE, German JB, et al. Extracellular fungal polyol lipids: A new class of potential high value lipids. Biotechnology Advances. 2018;**36**:397-414

[108] Mittler R, Vanderauwera S, Suzuki N, Miller G, Tognetti VB, Vandepoele K, et al. ROS signaling: The new wave? Trends in Plant Science. 2011;**16**(6):300-309. DOI: 10.1016/j.tplants.2011.03.007

[109] Sridharan R, Muralidharan M, Kumar PS, Gayathri KV, Rangasamy G. Extraction of biosurfactants from LDPE and PAH degrading bacterial strains isolated from plastics and oil contaminated sites: Statistical

optimization using response surface methodology. Journal of Hazardous Materials Advances. 2025;**17**:100603. DOI: 10.1016/j.hazadv.2025.100603

[110] Ambaye TG, Vaccari M, Prasad S, Rtimi S. Preparation, characterization and application of biosurfactant in various industries: A critical review on progress, challenges and perspectives. Environmental Technology & Innovation. 2021;**24**:102090. DOI: 10.1016/j.eti.2021.102090

[111] Hossain MS, Dietz K-J. Tuning of redox regulatory mechanisms, reactive oxygen species and redox homeostasis under salinity stress. Frontiers in Plant Science. 2016;**7**:548. DOI: 10.3389/fpls.2016.00548

[112] Bochynek M, Hodurek P, Łukaszewicz M, Lewińska A. Centrifugal partition chromatography as a potential method of isolation and purification of amphiphilic substances from a solid-state fermentation process. Biomass Conversion and Biorefinery. 2023;**13**(18):16333-16343. DOI: 10.1007/s13399-023-03821-w

[113] Borah SN, Sen S, Pakshirajan K. Biosurfactants for enhanced bioavailability of micronutrients in soil. In: Biosurfactants for a Sustainable Future. Hoboken, NJ, USA: Wiley; 2021. pp. 159-181. DOI: 10.1002/9781119671022.ch8

[114] Wang Y, Wang J, Sarwar R, Zhang W, Geng R, Zhu K-M, et al. Research progress on the physiological response and molecular mechanism of cold response in plants. Frontiers in Plant Science. 2024;**14**:1334913. DOI: 10.3389/fpls.2024.1334913

[115] Becker VI, Goessling JW, Duarte B, Caçador I, Liu F, Rosenqvist E, et al. Combined effects of soil salinity and high temperature on photosynthesis and growth of quinoa plants (*Chenopodium quinoa*). Functional Plant Biology. 2017;**44**(7):665. DOI: 10.1071/FP16370

[116] Elumalai P, Rajamohan R, Thashbin Vayal Purayil A, Menon V, Pranav Srivatsan R, Santhosh Kumar A, et al. Biosurfactant and iron oxide nanoparticle-assisted bioremediation of soil co-contaminated with hydrocarbons and hazardous heavy metals. Chemical Engineering Journal. 2024;**497**:154677. DOI: 10.1016/j.cej.2024.154677

[117] Soberón-Chávez G, Hausmann R, Maier RM, Déziel E. Challenges and prospects for microbial biosurfactant research. In: Biosurfactants. Amsterdam, Netherlands: Elsevier; 2023. pp. 305-311. DOI: 10.1016/B978-0-323-91697-4.00013-2

[118] Vieira IMM, Santos BLP, Ruzene DS, Silva DP. An overview of current research and developments in biosurfactants. Journal of Industrial and Engineering Chemistry. 2021;**100**:1-18. DOI: 10.1016/j.jiec.2021.05.017

[119] Iqbal MZ, Singh K, Chandra R. Recent advances of plant growth promoting rhizobacteria (PGPR) for eco-restoration of polluted soil. Cleaner Engineering and Technology. 2024;**23**:100845. DOI: 10.1016/j.clet.2024.100845

[120] Yasmin A, Aslam F, Fariq A. Genetic evidences of biosurfactant production in two *Bacillus subtilis* strains MB415 and MB418 isolated from oil contaminated soil. Frontiers in Bioengineering and Biotechnology. 2022;**10**:855762. DOI: 10.3389/fbioe.2022.855762

[121] Meištininkas R, Vaškevičienė I, Dikšaitytė A, Pedišius N, Žaltauskaitė J. Biosurfactant-assisted phytoremediation of diesel-contaminated soil by three

different legume species. Environments. 2024;**11**(4):64. DOI: 10.3390/environments11040064

[122] Gusain S, Joshi S, Joshi R. Sensing, signalling, and regulatory mechanism of cold-stress tolerance in plants. Plant Physiology and Biochemistry. 2023;**197**:107646. DOI: 10.1016/j.plaphy.2023.107646

[123] Vigneshwaran C, Vasantharaj K, Krishnanand N, Sivasubramanian V. Production optimization, purification and characterization of lipopeptide biosurfactant obtained from *Brevibacillus* sp. AVN13. Journal of Environmental Chemical Engineering. 2021;**9**(1):104867. DOI: 10.1016/j.jece.2020.104867

[124] Zaynab M, Fatima M, Sharif Y, Sughra K, Sajid M, Khan KA, et al. Health and environmental effects of silent killers organochlorine pesticides and polychlorinated biphenyl. Journal of King Saud University - Science. 2021;**33**(6):101511. DOI: 10.1016/j.jksus.2021.101511

[125] Etesami H, Santoyo G. Boosting rhizobium-legume symbiosis: The role of nodule non-rhizobial bacteria in hormonal and nutritional regulation under stress. Microbiological Research. 2025;**297**:128192. DOI: 10.1016/j.micres.2025.128192

[126] Brokate O, Papenbrock J, Turcios AE. Biofilm-forming microorganisms in the rhizosphere to improve plant growth: Coping with abiotic stress and environmental pollution. Applied Soil Ecology. 2024;**202**:105591. DOI: 10.1016/j.apsoil.2024.105591

[127] Silva IA, Fortunato JGLA, Almeida FCG, Alves RN, Cunha MCC, Rufino RD, et al. Production and application of a new biosurfactant for solubilisation and mobilisation of residual oil from sand and seawater. PRO. 2024;**12**(8):1605. DOI: 10.3390/pr12081605

[128] Sharma J, Sundar D, Srivastava P. Biosurfactants: Potential agents for controlling cellular communication, motility, and antagonism. Frontiers in Molecular Biosciences. 2021;**8**:727070. DOI: 10.3389/fmolb.2021.727070

[129] Hellweg T, Oberdisse J, Sottmann T. Recent advances in biosurfactant-based association colloids–formation of microemulsions. Frontiers in Soft Matter. 2023;**3**:1260211. DOI: 10.3389/frsfm.2023.1260211

[130] Schonhoff A, Stöckigt G, Wulf C, Zapp P, Kuckshinrichs W. Biosurfactants' production with substrates from the sugar industry – Environmental, cost, market, and social aspects. RSC Sustainability. 2023;**1**(7):1798-1813. DOI: 10.1039/D3SU00122A

[131] Hu X, Wang C, Wang P. Optimization and characterization of biosurfactant production from marine *Vibrio* sp. strain 3B-2. Frontiers in Microbiology. 2015;**6**:976

[132] Nagtode VS, Cardoza C, Yasin HKA, Mali SN, Tambe SM, Roy P, et al. Green surfactants (biosurfactants): A petroleum-free substitute for sustainability—comparison, applications, market, and future prospects. ACS Omega. 2023;**8**(13):11674-11699. DOI: 10.1021/acsomega.3c00591

[133] Narenkumar J, Das B, Abilaji S, Sathishkumar K, AlSalhi MS, Devanesan S, et al. Biosurfactant-assisted bio-electrokinetic enhanced remediation of heavy metal-contaminated soil. Frontiers in Microbiology. 2024;**15**:1458369. DOI: 10.3389/fmicb.2024.1458369

[134] Sunde M, Pham CLL, Kwan AH. Molecular characteristics and biological functions of surface-active and surfactant proteins. Annual Review of Biochemistry. 2017;**86**:585-608. DOI: 10.1146/annurev-biochem-061516-045303

[135] Selva Filho AAP, Converti A, de Soares da Silva RCF, Sarubbo LA. Biosurfactants as multifunctional remediation agents of environmental pollutants generated by the petroleum industry. Energies. 2023;**16**(3):1209. DOI: 10.3390/en16031209

[136] Tao Y, Qiao Q, Ruan Y, Fang X, Wang X, Zhang Y, et al. SIM imaging of bacterial membrane dynamics and lipid peroxidation during photodynamic inactivation with a dual-functional activatable probe. Chemical Science. 2025;**16**(18):7766-7772. DOI: 10.1039/D5SC00858A

[137] Datta D, Ghosh S, Kumar S, Gangola S, Majumdar B, Saha R, et al. Microbial biosurfactants: Multifarious applications in sustainable agriculture. Microbiological Research. 2024;**279**:127551. DOI: 10.1016/j.micres.2023.127551

[138] Crouzet J, Arguelles-Arias A, Dhondt-Cordelier S, Cordelier S, Pršić J, Hoff G, et al. Biosurfactants in plant protection against diseases: Rhamnolipids and lipopeptides case study. Frontiers in Bioengineering and Biotechnology. 2020;**8**:2772. DOI: 10.3389/fbioe.2020.01014

[139] Parthasarathi T, Firdous S, Mariya David E, Lesharadevi K, Djanaguiraman M. Effects of high temperature on crops. In: Advances in Plant Defense Mechanisms. London, UK: IntechOpen; 2022. DOI: 10.5772/intechopen.105945

[140] Chen H, Song Y, Li H, Zaman S, Fan K, Ding Z, et al. Enhancing the adaptability of tea plants (*Camellia sinensis* L.) to high-temperature stress with small peptides and biosurfactants. Plants. 2023;**12**(15):2817. DOI: 10.3390/plants12152817

[141] Balan SS, Kumar CG, Jayalakshmi S. Physicochemical, structural and biological evaluation of Cybersan (trigalactomargarate), a new glycolipid biosurfactant produced by a marine yeast, *Cyberlindnera saturnus* strain SBPN-27. Process Biochemistry. 2019;**80**:171-180. DOI: 10.1016/j.procbio.2019.02.005

[142] Nehra A, Kalwan G, Taneja D, Jangra R, Joshi K, Kumar A, et al. Comprehensive structural, evolutionary and functional analysis of superoxide dismutase gene family revealed critical role in salinity and drought stress responses in chickpea (*Cicer arietinum* L.). Plant Physiology and Biochemistry. 2025;**226**:110042. DOI: 10.1016/j.plaphy.2025.110042

[143] Gong Z, Xiong L, Shi H, Yang S, Herrera-Estrella LR, Xu G, et al. Plant abiotic stress response and nutrient use efficiency. Science China Life Sciences. 2020;**63**(5):635-674. DOI: 10.1007/s11427-020-1683-x

[144] Bhatia S, Singh A, Batra N, Singh J. Microbial production and biotechnological applications of α-galactosidase. International Journal of Biological Macromolecules. 2020;**150**:1294-1313. DOI: 10.1016/j.ijbiomac.2019.10.140

[145] Zhu J-K. Abiotic stress Signaling and responses in plants. Cell. 2016;**167**(2):313-324. DOI: 10.1016/j.cell.2016.08.029

[146] Mitra S, Saran RK, Srivastava S, Rensing C. Pesticides in the

environment: Degradation routes, pesticide transformation products and ecotoxicological considerations. Science of the Total Environment. 2024;**935**:173026. DOI: 10.1016/j.scitotenv.2024.173026

[147] Zhang X, Köster P, Schlücking K, Balcerowicz D, Hashimoto K, Kuchitsu K, et al. CBL1-CIPK26-mediated phosphorylation enhances activity of the NADPH oxidase RBOHC, but is dispensable for root hair growth. FEBS Letters. 2018;**592**(15):2582-2593. DOI: 10.1002/1873-3468.13187

[148] Ali N, Wang F, Xu B, Safdar B, Ullah A, Naveed M, et al. Production and application of biosurfactant produced by *Bacillus licheniformis* Ali5 in enhanced oil recovery and motor oil removal from contaminated sand. Molecules. 2019;**24**(24):4448. DOI: 10.3390/molecules24244448

[149] Alkharabsheh HM, Seleiman MF, Hewedy OA, Battaglia ML, Jalal RS, Alhammad BA, et al. Field crop responses and management strategies to mitigate soil salinity in modern agriculture: A review. Agronomy. 2021;**11**(11):2299. DOI: 10.3390/agronomy11112299

[150] Hönig M, Roeber VM, Schmülling T, Cortleven A. Chemical priming of plant defense responses to pathogen attacks. Frontiers in Plant Science. 2023;**14**:1146577. DOI: 10.3389/fpls.2023.1146577

[151] Zhang J, Jia W, Yang J, Ismail AM. Role of ABA in integrating plant responses to drought and salt stresses. Field Crops Research. 2006;**97**(1):111-119. DOI: 10.1016/j.fcr.2005.08.018

[152] Mirzaei Aminiyan M, Shorafa M, Pourbabaee AA. Mitigating the detrimental impacts of low- and high-density polyethylene microplastics using a novel microbial consortium on a soil-plant system: Insights and interactions. Ecotoxicology and Environmental Safety. 2024;**283**:116805. DOI: 10.1016/j.ecoenv.2024.116805

[153] Mittler R. Oxidative stress, antioxidants and stress tolerance. Trends in Plant Science. 2002;**7**(9):405-410. DOI: 10.1016/S1360-1385(02)02312-9

[154] Araújo J, Rocha J, Oliveira Filho M, Matias S, Júnior SO, Padilha C. Rhamnolipids biosurfactants from Pseudomonas aeruginosa – A review. Biosciences Biotechnology Research Asia. 2018;**15**:767-781. DOI: 10.13005/bbra/2685

[155] Schultz J, Rosado AS. Extreme environments: A source of biosurfactants for biotechnological applications. Extremophiles. 2020;**24**:189-206. DOI: 10.1007/s00792-019-01151-2

[156] Chen C, Li D, Li R, Shen F, Xiao G, Zhou J. Enhanced biosurfactant production in a continuous fermentation coupled with in situ foam separation. Chemical Engineering and Processing - Process Intensification. 2021;**159**:108206. DOI: 10.1016/j.cep.2020.108206

[157] Habib S, Ahmed A. Expression profiling of stress protectants in *Zea mays* L. plants treated with *Bacillus pumilus*: Surfactant-producing bioeffector. International journal of Environmental Science and Technology. 2023;**20**(4):4435-4446. DOI: 10.1007/s13762-022-04089-9

[158] da Rocha Junior RB, Meira HM, Almeida DG, Rufino RD, Luna JM, Santos VA, et al. Application of a low-cost biosurfactant in heavy metal remediation processes. Biodegradation. 2019;**30**:215-233. DOI: 10.1007/s10532-018-9833-1

[159] Yang J, Duan G, Li C, Liu L, Han G, Zhang Y, et al. The crosstalks between Jasmonic acid and other plant hormone signaling highlight the involvement of Jasmonic acid as a core component in plant response to biotic and abiotic stresses. Frontiers in Plant Science. 2019;**10**:1349. DOI: 10.3389/fpls.2019.01349

[160] Jańczak-Pieniążek M, Cichoński J, Michalik P, Chrzanowski G. Effect of heavy metal stress on phenolic compounds accumulation in winter wheat plants. Molecules. 2022;**28**(1):241. DOI: 10.3390/molecules28010241

[161] Wang Z, Yu Y, Chen X, Li Y, Jones A, Rose RJ, et al. Mitigating drought-associated reproductive failure in maize: From physiological mechanisms to practical solutions. The Crop Journal. 2025;**13**(4):1022-1031. DOI: 10.1016/j.cj.2025.05.004

[162] Cohen J, Agel L, Barlow M, Garfinkel CI, White I. Linking Arctic variability and change with extreme winter weather in the United States. Science. 2021;**373**(6559):1116-1121. DOI: 10.1126/science.abi9167

[163] de Procópio SO, Barizon RRM, Pazianotto RAA, Morandi MAB, Braz GBP. Impacts of weed resistance to glyphosate on herbicide commercialization in Brazil. Agriculture. 2024;**14**(12):2315. DOI: 10.3390/agriculture14122315

[164] Guo J, Liu S, Li X, Liu F. Crop exposure to cold stress: Responses in physiological, biochemical and molecular levels. In: Sustainable Crop Productivity and Quality under Climate Change. Elsevier; 2022. pp. 1-19. DOI: 10.1016/B978-0-323-85449-8.00011-7

[165] Singh M, Singh SK, Sharma JG, Giri B. Insights into the multifaceted roles of soil microbes in mitigating abiotic stress in crop plants: A review. Environmental and Experimental Botany. 2024;**228**:106010. DOI: 10.1016/j.envexpbot.2024.106010

[166] Waqar A, Othman I, Shafiq N. The role of sustainable procurement in attaining sustainable development goals: Insights from sustainable logistics and sustainable metamorphosis practices. Journal of Cleaner Production. 2025;**500**:145285. DOI: 10.1016/j.jclepro.2025.145285

[167] Othman AR, Ismail NS, Abdullah SRS, Hasan HA, Kurniawan SB, Sharuddin SSN, et al. Potential of indigenous biosurfactant-producing fungi from real crude oil sludge in total petroleum hydrocarbon degradation and its future research prospects. Journal of Environmental Chemical Engineering. 2022;**10**(3):107621. DOI: 10.1016/j.jece.2022.107621

[168] Pierre E, Shaw E, Corr B, Pageau K, Rippa S. Applications of rhamnolipid biosurfactants in agriculture. Plant Stress. 2025;**15**:100749. DOI: 10.1016/j.stress.2025.100749

[169] Sankhyan S, Kumar P, Pandit S, Kumar S, Ranjan N, Ray S. Biological machinery for the production of biosurfactant and their potential applications. Microbiological Research. 2024;**285**:127765. DOI: 10.1016/j.micres.2024.127765

[170] Li Y, Narayanan M, Shi X, Chen X, Li Z, Ma Y. Biofilms formation in plant growth-promoting bacteria for alleviating agro-environmental stress. Science of the Total Environment. 2024;**907**:167774. DOI: 10.1016/j.scitotenv.2023.167774

[171] Goswami M, Deka S. Biosurfactant production by a rhizosphere bacteria

Bacillus altitudinis MS16 and its promising emulsification and antifungal activity. Colloids and Surfaces B: Biointerfaces. 2019;**178**:285-296. DOI: 10.1016/j.colsurfb.2019.03.003

[172] Bamal D, Duhan A, Pal A, Beniwal RK, Kumawat P, Dhanda S, et al. Herbicide risks to non-target species and the environment: A review. Environmental Chemistry Letters. 2024;**22**(6):2977-3032. DOI: 10.1007/s10311-024-01773-9

[173] Luengo GS, Fameau A-L, Léonforte F, Greaves AJ. Surface science of cosmetic substrates, cleansing actives and formulations. Advances in Colloid and Interface Science. 2021;**290**:102383. DOI: 10.1016/j.cis.2021.102383

[174] Sanchez-Huerta C, Zhang S, Alahmari M, Humam AA, Hong P-Y. Remediation of petroleum hydrocarbons in contaminated groundwater with the use of surfactants and biosurfactants. Chemosphere. 2025;**376**:144290. DOI: 10.1016/j.chemosphere.2025.144290

[175] Adetunji AI, Olaniran AO. Production and potential biotechnological applications of microbial surfactants: An overview. Saudi Journal of Biological Sciences. 2021;**28**:669-679. DOI: 10.1016/j.sjbs.2020.11.005

[176] Wang Z, Li G, Sun H, Ma L, Guo Y, Zhao Z, et al. Effects of drought stress on photosynthesis and photosynthetic electron transport chain in young apple tree leaves. Biology Open. 2018;**7**(11):bio035279. DOI: 10.1242/bio.035279

[177] Abdelfattah F, Salah M, Dahleez K, Darwazeh R, Halbusi HA. Public policy and sustainability: How green core competence, government trust, and policy satisfaction influence green R&D investments in the private sector. Sustainable Futures. 2025;**9**:100461. DOI: 10.1016/j.sftr.2025.100461

[178] Abdoli S, Asgari Lajayer B, Bagheri Novair S, Price GW. Unlocking the potential of biosurfactants in agriculture: Novel applications and future directions. Sustainability. 2025;**17**(5):2110. DOI: 10.3390/su17052110

[179] Venkataraman S, Rajendran DS, Kumar PS, Vo D-VN, Vaidyanathan VK. Extraction, purification and applications of biosurfactants based on microbial-derived glycolipids and lipopeptides: A review. Environmental Chemistry Letters. 2022;**20**(1):949-970. DOI: 10.1007/s10311-021-01336-2

Chapter 6

Serendipita indica: A Plant Elixir for Growth Enhancement and Environmental Stress Resilience

Sana Saleem

Abstract

Serendipita indica is an endophytic fungus that has received widespread interest as a potential solution for alleviating the effects of climate change on agriculture and minimizing dependency on chemical fertilizers. Furthermore, it has characteristic growth-enhancing and stress-reducing properties, which are becoming an important element of sustainable agricultural techniques. This fungus' capacity to colonize many plant hosts while improving nutrient absorption, biomass output, and root architecture makes it a viable biofertilizer. Unlike AMF, *S. indica* may be cultivated relatively easily, allowing for large-scale usage in agricultural techniques. More significantly, *S. indica's* superior morphophysiological and biochemical systems help to tolerate abiotic challenges such as drought, salt, temperature changes, and heavy metal toxicity. Furthermore, its capacity to stimulate nutrient absorption, improve root system architecture, and increase biomass output provides a natural alternative to chemical fertilizers, therefore decreasing the negative environmental effects of contemporary farming techniques. These utilize a number of complicated strategies, including the management of phytohormonal pathways, activation of stress-sensitive genes, increase in antioxidant enzyme activity, and maintaining ion homeostasis, all of which improve plant resilience under adverse conditions. This review provides a comprehensive overview of *S. indica* as a growth promoter and abiotic stress modulator and its potential for transforming modern agriculture and overcoming the challenges posed by global climate change and reducing the dependency on chemical fertilizers.

Keywords: beneficial microorganisms, sustainable agriculture, biocontrol agent, symbiotic relationship, plant growth promotion

1. Introduction

The effectiveness of PGPMs as biofertilizers has been emphasized by researchers, pointing out their ability to improve the availability of key nutrients such as nitrogen, phosphorus, potassium, and others, along with modulating phytohormonal mechanisms [1]. According to Lobo et al. and Etesami [2, 3], microorganisms play a significant role in improving soil and plant nutrition, regulating phytohormones, mitigating biotic and abiotic stresses, and promoting plant growth and productivity. A diverse array of advantageous organisms inhabits the root zone of plants, including various bacterial and fungal groups that colonize and support their host plants, collectively referred to as plant growth promoting microbes (PGPM) [4]. PGPMs contribute to plant growth, development, phytoremediation potential, and enhanced stress tolerance through various physiological mechanisms. The interactions between plants and microbes establish mutually beneficial relationships, with microbes playing a vital role in helping plants adapt to toxic environments and promoting their growth, thereby improving phytoremediation efficiency and tolerance to abiotic stresses [5]. These microorganisms also play a pivotal role in mitigating plant diseases by inducing resistance against various pathogens [6]. Furthermore, PGPMs enhance the biosynthesis of key phytohormones, including auxins, cytokinin, abscisic acid, ethylene, brassinosteroids, jasmonic acid, salicylic acid, strigolactones, and gibberellins, which collectively stimulate plant growth and fortify plants against diverse stress factors [7]. A notable advantage of PGPM is their contribution to bioremediation, effectively reducing heavy metal toxicity in contaminated soils [8]. Furthermore, incorporating PGPM into hydroponics systems can enhance the quality of recycled water by decomposing organic matter, mitigating the accumulation of harmful substances, and stabilizing pH levels. This encourages the effective and prolonged reuse of reclaimed water, eliminating the need for frequent refills and guaranteeing long-term resource efficiency [9].

PGPMs encompass endophytic fungi, which have become a focal point in research due to their unique abilities to promote plant growth and enhance resilience to both biotic and abiotic stresses. Arbuscular mycorrhizal fungi (AMF) stand out as one of the most critical beneficial fungi, known for its growth promoting effects and capacity to strengthen plant defense systems [10]. However, AMF is limited by host specificity, which restricts its application to certain plant species. Moreover, its cultivation is dependent on specific host plants, leading to increased production costs, extended time requirements, and the necessity for substantial space allocation [11]. *Serendipita indica*, a fascinating endophytic fungus, has great potential for both plant growth promotion and disease management. It is gaining more scientific interest because of its unique ability to overcome the constraints of AMF. Consequently, this chapter focuses on the research of *S. indica* as a plant defense mechanism against various abiotic stressors and its potential as a growth promoting agent.

2. *S. indica* and its cultivation

According to molecular phylogenetic study, *Serendipita indica*, a filamentous fungus that colonizes a variety of plants and has a symbiotic connection with them [12]. This fungus is a member of the family Serendipitaceae and order Sebacinales within the Basidiomycota division [13]. It was later moved to the genus *Serendipita* from its prior name, *Piriformospora indica*. It was discovered by Varma in rhizospheres of *Prosopis julifora* and *Zizyphus nummularia*, which were grown in the Indian Thar Desert [14]. *S. indica* does not require host plants to proliferate and therefore can be axenically cultivated on a variety of complicated and minimum substrates. For axenic *S. indica* propagation, a variety of media have been proposed; however, these media are either simple and cost-effective but less effective or they are complicated, require certain specifications, and can be laborious to manufacture [15, 16]. Additionally, it may be cultivated on modified Melin-Norkrans (MMN) media, potato dextrose agar

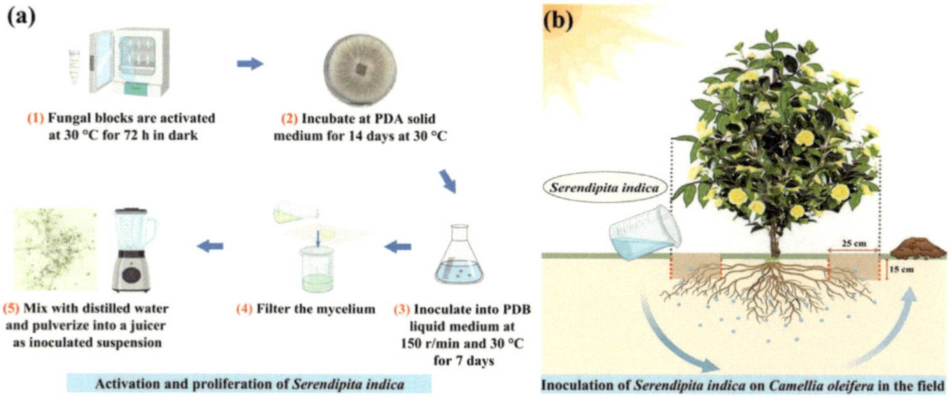

Figure 1.
Preparation and application of S. indica *[20].*

(PDA) or broth (PDB), aspergillus medium, modified aspergillus medium, and malt extract [17]. It was also reported by Osman et al. [16] that, for better *S. indica* axenical development, several CM media changes have been proposed, utilizing various sources and concentrations of carbohydrates, and optimizing the pace of agitation for solution. Nevertheless, it has been observed that optimized vegetable juice (VJ) medium is better than complicated media that are frequently employed to produce fungus *S. indica*.

Growing it entails using any media until a significant volume of mycelium is developed. Pore extraction and optimization, or blending the obtained culture using a blender, can be employed. This is followed by the preparation of suspensions or inoculants, depending on the intended application method. Concentration optimization is crucial, with 1–3% (w/v) demonstrating effective results on host plants. Higher concentrations did not yield significant additional benefits [18]. The created inoculum can be used to treat seeds, roots, and can be used as substrate application [19]. **Figure 1** shows preparation and inoculation of *S. indica* [20].

3. Colonization of *Serendipita indica* as endophytic fungi

Serendipita indica has emerged as a focal point of interest due to its exceptional traits, including broad host-plant compatibility and adaptability to various synthetic growth media. This fungus significantly enhances the growth and development of diverse plant species while bolstering their resilience against a wide spectrum of biotic and abiotic challenges [21, 22]. *S. indica* uses a colonization method that consists of a biotrophic relationship at the beginning and a growth phase that is linked to controlled host cell loss, which does not lessen the positive effects on the plant host. One of the most important aspects of *S. indica* 's colonization approach is the activation of programmed cell death (PCD) in the epidermal and cortical layers, which is followed by a surge in the synthesis of beneficial enzymes [23]. Charura et al. [23] further emphasize the significance of PCD, which is governed by the downregulation of BFN1 gene expression. This illustrates the transcriptional regulatory mechanism that makes it easier for beneficial microorganisms to reside in the roots. In addition, it is believed that effective colonization is influenced by the metabolism of carbohydrates and the regulation of many genes linked to plant defense and hormone signaling [24].

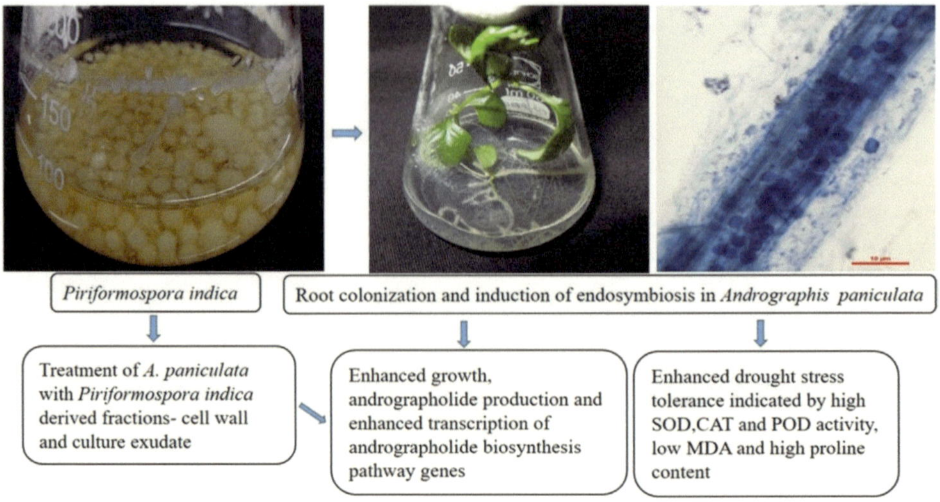

Figure 2.
S. indica *cultivation, colonization of* S. indica *and its effect on host plant [30].*

S. indica enters the cell wall after it has been degraded by mechanical pressure and several enzymes including cellulase, protease, lipase, xylanase, chitinase, amylase, and pectinase [25]. This endophytic fungus produces hyphae and chlamydospores, which resemble beaded chains [16].

According to Gill et al. [26] hyphae are hyaline have thin walls and have an irregular septum that contains many nuclei. Chlamydospores are asexual, pear-shaped spores that can be found alone or in clusters. The hyphae are wildly entangled and resemble tangled branches and ropes. Various proteins and polysaccharides envelop the hyphal walls [27].

Upon colonization with host plants, *S. indica* exhibits a variety of desirable potential that are highly relevant to contemporary agriculture. It has been shown that *S. indica* might assist plants tolerate stress and overcome severe environmental factors including salt, drought, and heavy metal toxicity. It has positive impacts on host plant output and seed characteristics in both normal and stressful situations [28, 29]. **Figure 2** illustrates the colonization of *S. indica* with roots of host plant and highlights its beneficial effects on the host plant [30].

4. Growth enhancement in host plants by *S. indica*

4.1 Growth enhancement by nutrient uptake

The connection between endophytic fungus and plants can benefit the latter in a number of ways, including higher intake of mineral elements such as phosphorus and potassium, sulfur, and magnesium [31]. Besides, these fungi can also benefit from root exudation of host plants [32]. It was demonstrated that, instead of phosphorus being transferred from the roots to the leaves, *S. indica* would have been engaged in the acquisition of root phosphorus, as evidenced by the rise in phosphorus concentrations in the roots opposed to those in the leaves [20]. Therefore, nutrient absorption is improved by colonization with *S. indica*. A rise in the movement of nutrients within the plant cell may be the cause of this phenomena. Besides, member of the phosphate: H1 symporter (PHS) family, the high-affinity phosphate (PO_4^{3-}) transporter (*PiPT*), is found in fungi. The phosphate intake of plants is said to be enhanced by this *S. indica* transporter [33]. Moreover, existence of numerous K^+ transporters in *S. indica*, including *SiHAK1, SiTRK1, SiTRK2*, and *SiTOK1* have been discovered [34]. Remarkably, by expressing its own K+ transporter gene SiHAK1, *S. indica* improves its own development while decreasing the host plant's total K^+ accumulation when there is a K^+ shortfall. However, *S. indica* did not stimulate host plant growth in the low K^+ environment [35]. Furthermore, *S. indica* has been shown to promote development by stimulating nitrogen build up in the roots of tobacco and Arabidopsis seedlings through the induction of nitrate reductase and the starch-degrading enzyme glucan-water dikinase (*SEX1*) [36]. Additionally, *S. indica* has been shown to enhance tomato colonization's nitrogen uptake, which promotes the growth and development of the host plant [37].

The effect of *S. indica* on the intake of key elements like magnesium and sulfur has been investigated, resulting in increased plant growth. A magnesium transporter termed *PiMgT1*, member of CorA-like protein family of bacteria, was identified and functionally described in *S. indica* [38]. This could be mechanism underlying enhanced Mg^{2+} uptake in the host plant. Moreover, a sulfur-specific high-affinity transporter, *SiSulT*, has been discovered and described in *S. indica* during its interaction with maize [39]. Under low-sulfate circumstances, this transporter has been shown to be essential for the absorption of sulfur in the sulfate form (SO_4^{2-}) to the host plant and for fostering the host's growth and development. Thus, in the mutualistic interplay between plants and endophytic fungi, the fungi utilize root exudates as a nutrient source and, in return, provide significant benefits to the host [40]. Additionally, *S. indica* enhances chlorophyll synthesis, which drives an increased photosynthetic rate. This process facilitates the accumulation of essential leaf carbohydrates such as glucose, fructose, and sucrose [20].

4.2 Elevation of plant growth by phytohormones

During colonization, *S. Indica* modulates a number of phytohormonal-regulated pathways, including the glucosinolate pathway in the early phases of contact and the jasmonate-regulated secondary metabolite [41]. Regulation and production of auxins and gibberellins for recognition process, in the initial process of colonization, changes to auxin transport that *S. Indica* induces in its host plant are a crucial component of the growth-promoting impact [34]. Auxin exporter gene *PIN2* expression is particularly suppressed by the colonizing fungus during infection of host roots, which also stimulates auxin signaling pathways. Root tips accumulate auxin as a result of this repression, which lowers basipetal auxin transport. Plant biomass is strongly correlated with a decrease in *PIN2* quantity or functioning, as seen in roots infected with *S. indica* or the agr1–1 mutant [42]. Furthermore, in tomato, *S. indica* promotes putrescine production through the route regulated by *ARGININE DECARBOXYLASE 1 (SlADC1).* Gene expression research and metabolomic analysis show that *S. indica* colonization raises tomato putrescine content and *SlADC1* transcript levels. Furthermore, auxin (indole-3-acetic acid) and some gibberellins (GA4 and GA7) are increased in tomatoes when putrescine is applied exogenously. This suggests that *S. Indica* uses putrescine to increase growth-promoting phytohormones, which in turn leads to improved plant growth [34].

GA signaling is enlisted by *S. indica* to initiate colonization of root cells. A higher GA level in the colonized roots was linked to the growth promotion of Chinese cabbage and barley seedlings by *S. indica,* while downregulated GA2ox gene in the roots of host plant, responsible for inactivating GA, was reported [43]. The flowering-repressing gene *FLOWERING LOCUS C* was down-regulated, but the transcript levels of the flowering regulatory genes *FLOWERING LOCUS T, LEAFY,* and *APETALA1* as well as the GA biosynthesis genes *gibberellin 20-oxidase 2, gibberellin 3-oxidase* 1, and gibberellin requiring 1 were higher in colonized plants. The analysis of GA content demonstrated that *S. indica* colonization considerably elevated GA4 levels. In Arabidopsis, inoculation of the ga5 mutant, which lacks a GA biosynthetic gene, resulted in a less strong activation of flowering-related gene expression and a lesser rise in GA4 content than the wild-type. These results suggest that *S. indica* promotes early blooming in Arabidopsis via increasing GA levels [44]. During the initial detection phase preceding interaction between the two symbiotic partners, Arabidopsis roots and shoots showed a significant increase in abscisic acid (ABA) accumulation upon *S. indica* colonization [45]. Given its effect on host innate immunity, ABA enhances the interaction between Arabidopsis roots and *S. indica.* This will also have a significant influence on the formation and outcome of the symbiosis under stress [43].

5. Plant abiotic stress management by *S. indica*

The world's food supply is in jeopardy because abiotic stress has reduced agricultural lands and crop yields. For instance, 45% of the world's arable land is affected by soil dryness, and soil salinity can reduce agricultural production by 10–50% [46]. Extreme precipitation, high temperatures, and protracted drought expose crop development to abiotic stressors. Crop development and yield are further hindered by human-induced heavy metal pollution of the soil. Therefore, enhancing plants' ability to respond to abiotic stressors is essential. According to Yaghoubian et al. [47], *S. indica* can adapt to unfavorable conditions including high temperatures, high pH levels, a lack of nutrients in the soil, and high concentrations of heavy metals. Thus, it helps plants better withstand abiotic stressors such drought, low temperature, and salt stress [17, 48].

5.1 Drought stress

Drought is the one of the major threats from abiotic stress. Crop development and production are decreased by drought stress, which also interferes with physiological and biochemical functions such photosynthesis, transpiration, cell division, and chlorophyll concentration [48, 49]. Nonetheless, it has been shown that *S. indica* root colonization diminishes drought damage to host plants. It is widely acknowledged that *S. Indica* increases the growth of host plants and aids in better nutrient absorption, including nitrogen as well as improved water uptake by plants [49]. These positive effects are also seen under drought stress and are attributed to the origin of *S. indica* from arid region [25]. Besides, *S. indica*'s active drought adaptation was further evidenced by the significant rise in colonization rate in roots caused by soil dehydration, which was followed by a surge in the quantity chlamydospores [49]. Additionally, it could modify drought-avoidance strategies by improving the host plant's morphological and physiological characteristics. Expanding the root systems of plants colonized by *S. indica* has been found to promote nutrition and water intake, which in turn increases plant biomass, regulates leaf temperature, decreases leaf wilting, and enhances stomatal closure [50]. Additional benefits of *S. indica* include lowering photosynthetic efficiency and avoiding the breakdown of thylakoid and chlorophyll proteins, which lowers the danger of drought stress [51]. Studies have also demonstrated that *S. indica* raises catalase (CAT), superoxide dismutase (SOD), and ascorbate peroxidase (APX) and controls the genes associated with these compounds in plants [52, 53]. Other research, however, has documented a decline in host plant APX and other stress indicators, thus indicating that *S. indica* responds differently to antioxidant activity [54]. Plants inoculated with *S. indica* under drought stress had higher proline contents and less formation of malondialdehyde (MDA) [55]. Additionally, *S. Indica* regulates the activity of genes linked to proline and total antioxidant capacity (TAC), which improves the plant's resilience to drought, and enhances proline accumulation [51]. Hormonal regulation by hormonal modulation by *S. indica* is also shown, with plants treated with this fungus exhibiting higher ABA regulation. ABA significantly contributes to plants' increased ability to withstand drought stress [56].

5.2 Salinity stress

Inadequate methods of farming, extreme temperatures, excessive evapotranspiration, insufficient precipitation, and poor soil management all contribute to rising salinity in cropping soils [57], which has a significant effect on agricultural output. Salinity is recognized as a major abiotic stress, contributing to approximately 20% of damage to irrigated land. It is projected to impact nearly 50% of agricultural land by 2050 [58, 59]. The ionic and osmotic equilibrium in plants is altered by increased soil salinity, which is a serious hazard that may eventually cause plant death [60]. Plants undergo metabolic and physiological changes as a result of this disturbance [57]. High sodium (Na^+) concentrations drastically change the structure of soil by decreasing water conductivity, aeration, and soil porosity. Consequently, this reduces the water potential of the soil, which makes it more difficult for plants to absorb water and vital nutrients [61].

To mitigate the adverse effects of salinity on plants, the use of beneficial microorganisms is regarded as a promising and adaptable approach [62], as they are affordable, provide results quickly, and are environmentally sustainable [63]. It has been demonstrated that *S. indica,* a well-known soil microbe, increases plant resilience to salt stress [64]. Arabidopsis, barley, *Medicago truncatula*, rice, tomato, and wheat are among the many plant species that fall within this category [65, 66]. *S. indica* not only causes morphological and physiological changes in plants under salt stress but also significantly alters a number of molecular processes to increase plant resistance to salinity. Under salt stress, *S. indica* benefits plants by improving their photosynthetic state [58], increasing photosynthetic pigments [66] and total chlorophyll content (chlorophyll a and b) [59, 61], and improving photosystem II's photochemical efficiency [58]. As a result, colonized plants under salt stress have higher biomass, root and shoot branching, root and shoot length, and fresh and dry weight [58, 59]. Numerous investigations have demonstrated that, in saline soil, *S. indica* promotes K^+ absorption while inhibiting Na^+ uptake by plants [65, 67, 68]. Furthermore, plants infected with *S. indica* exhibit changed regulation of phytohormones such as ABA, which is elevated and helps to improve defensive systems [69]. Additionally, *S. indica* boosts plant hormone levels, including gibberellic acid (GA), proline, glycine, betaine, and salicylic acid (SA), which strengthens the plant's resilience to salinity stress [70, 71]. Enhanced oxidative stress tolerance is a result of the modulation of antioxidant mechanisms in the roots and shoots of *S. indica*-colonized plants exposed to salt stress. These mechanisms include regulation of peroxidase (POD), SOD, CAT, APX, glutathione reductase (GR), dehydroascorbate reductase (DHAR), monodehydroascorbate reductase (MDHAR), glutathion (GSH), flavonoids, and phenols [59, 72]. *S. indica* simultaneously lowers MDA and H_2O_2 levels in a number of plant tissues [72]. It was also observed that colonized plants under salt stress had altered gene regulation. Salinity stress causes increases in salt overly sensitive (SOS) and sodium/hydrogen antiporter (NHX) gene expression in plants colonized by *S. indica* [57]. Additionally, by boosting the expression of TIP and PIP family proteins, which control ionic homeostasis, water status, and nutrients, *S. indica* boosts plant resilience to salinity [58].

5.3 Water logging stress

Another environmental stress includes water logging stress. Waterlogging causes hypoxic conditions for plant roots by adversely affecting the physicochemical properties of the soil, such as pH, oxygen availability, and redox potential. Plant growth, development, and survival are hindered by these circumstances [73]. *S. indica* has been seen to alleviate the effect of water logging. According to Liang et al. [74], waterlogging treatments promote *S. indica* colonization in host roots, which results in the build-up of chlamydospores inside the roots. These chlamydospores serve as structures for waterlogged roots to retain oxygen, reducing the stress conditions brought on by water logging [75]. Besides, *S. indica* enhances protein, vitamin C, total soluble sugar, and chlorophyll contents in Chinese cabbage, thus increasing the yield [76]. By controlling indoleacetic acid (IAA) levels and raising root proline content, *S. indica*-inoculated plants demonstrated enhanced growth under waterlogging circumstances. This was followed by increased leaf water potential and increased expression of the Δ1-pyrroline-5-carboxylate synthetase 2 (P5CS2) gene [77]. According to these results, S. indica could potentially be used as a biological stimulant to increase host plants' resistance to waterlogging.

5.4 Low-temperature stress

The changing climate worldwide has accelerated in recent years, leading to an increase in severe weather events, such as future occurrences of exceptionally low temperatures [78]. Plants are most harmed by low temperatures at the level of their cellular membranes, which interfere with selective permeability and accelerate the loss of electrolytes like K^+ and amino acids [79]. Consequently, the growth, development, and yield of plants are adversely affected [80]. *S. indica's* resilience to low temperatures is demonstrated by its capacity to thrive in plants at temperatures ranging from −30 to 4°C [81]. It has been demonstrated that *S. indica* colonization increases soluble sugar levels and stimulates the synthesis of new proteins, which helps the plants adapt to cold stress [76]. Furthermore, Shen and Chen [81] found that low temperatures significantly decreased biomass production, factors related to leaf gas exchange, and root fungal colonization. At low temperatures, however, *S. indica* inoculation significantly raised the biomass of shoots and roots, photosynthetic rate, stomatal conductance, and transpiration rate. Proline levels and the activities of CAT, SOD in leaves under low-temperature stress were also markedly increased by *S. indica* inoculation. In order to further support its function as a strong microbial biostimulant to boost low-temperature tolerance, *S. indica* additionally increased the expression of the *PcRS* and *PcRS11* genes. Additionally, by upregulating the expression of the *CBF* and *COR* genes, *S. indica* colonization has improved tolerance to low temperatures and sped up recovery [78]. In frozen *Arabidopsis* seedlings, *S. indica* also increased important transcription factors as *WRKY, ERF*, and *MYB* [82], thus significantly bolstering the beneficial effects of *S. indica* under cold stress.

5.5 Heavy metal stress

Globally, heavy metal contamination in agro systems has grown worse during the past few decades. According to Feng et al. [83], heavy metals are frequently carried into roots by certain transporter proteins across the plasma membrane and subsequently to the xylem or phloem via coplanar or plastid extracellular channels. According to Liu et al. [84], heavy metals alter protein structures, disturb the structure of membranes, and inactivate and denaturate intracellular enzymes in plants. These effects have an impact on regular metabolic activities in plants, including respiration, transpiration, and photosynthesis. Consequently, the redox equilibrium becomes perturbed [85]. *S. indica*'s distinct physiochemical characteristics make it an effective means of boosting plant tolerance by controlling gene expression in response to heavy metal stress. Heavy metal toxicity is decreased by *S. indica,* which also alters the plant's morphophysiological and biochemical processes, metabolic rate, and nutrient absorption [86]. Under Cadmium(Cd), Chromium(Cr), Lead(Pb) stress, *S. indica* inoculation increased the expression of chitinase genes and pathogenesis-related protein genes in tobacco [76]. In plants under Cd stress, *S. indica* increased the amounts of chlorophyll a, chlorophyll b, and total [87]. Likewise, in rice under Cd stress, *S. indica* downregulated the export-type transporter gene OsHMA2 and elevated the expression of OsHMA3, a Cd-specific protein gene implicated in vacuolar compartmentalization. This decreased the spread of Cd in the rice's aboveground portions [13]. To put it briefly, *S. indica* inoculation can cause physiological and molecular reactions in the host plant that is subjected to abiotic stressors, hence increasing the plants' resistance to such stresses.

6. Large scale production of *S. indica* for field application

Reduced use of artificial fertilizers and pesticides is necessary for sustainable agriculture, which makes fungi that promote plant development, like *S. indica,* particularly interesting [88]. *S. indica* is appealing for agroforestry, horticulture, and medicinal plants because of its quick and extensive in vitro multiplication, which enables its inoculation in the field [89]. *S. indica* inoculation into the host plant's rhizosphere under field circumstances dramatically elevated SOD, POD, and CAT activity in leaves, raised ascorbic acid and glutathione levels, and amplified the expression of POD, CAT1, and FAD6 genes [90]. Applying *S. indica* to peppermint (*Mentha piperita*) reduced the negative effects of salt stress by increasing essential oil components, including menthone, boosting antioxidant capacity, and decreasing membrane electrolyte leakage, as shown in [91]. Additionally, *S. indica* applied in the field reduces the disease incidence in barley caused by *Pseudocercosporella herpotrichoides* by inducing systemic resistance. Concurrently, the number of sheath layers increased [92]. Moreover, a formulation for field application has been prepared and named as "Rootonic," by mixing the biomass of *S. indica* and magnesium sulfite [93]. However, a lot of research is yet to be performed for the preparation of commercial biofertilizer. Additionally, by combining *S. indica* biomass with magnesium sulfite, a formulation known as "Rootonic" has been created for field application [93]. However, a substantial amount of research is still needed to prepare commercial biofertilizer.

7. Conclusion

S. indica has the potential to play a significant role in modern agriculture by mitigating the effects of increasing environmental stresses. As evidenced in various studies, this endophytic fungus enhances the tolerance of plants to both biotic and abiotic stresses. Its potential use as a biofertilizer is particularly high given its ability to colonize a wide range of plant species, its relatively easy culture compared to AMF, and its positive effects on plant growth and development. Nevertheless, while the benefits of *S. indica* are well documented, many aspects of its responses to abiotic and biotic stresses are still not clear. Therefore, large gaps exist regarding its complex molecular interaction with plants, particularly in multifaceted and evolving environmental conditions. The progress of research that applies omics technologies, including transcriptomics, proteomics, and metabolomics, is needed to unravel the complex functions related to plant resilience. Additionally, there should be more initiatives undertaken to commercialize this beneficial product. This will be essential to maximize its use in sustainable agricultural practices.

Conflict of interest

The authors declare no conflict of interest.

Author details

Sana Saleem
Mendel University, Brno, Czech Republic

*Address all correspondence to: sanasaleemsg009@gmail.com; xsaleem@mendelu.cz

References

[1] Laishram B, Devi OR, Dutta R, et al. Plant-microbe interactions: PGPM as microbial inoculants/biofertilizers for sustaining crop productivity and soil fertility. Current Research in Microbial Sciences. 2024;**8**:100333. DOI: 10.1016/j.crmicr.2024.100333

[2] Lobo CB, Juárez Tomás MS, Viruel E, et al. Development of low-cost formulations of plant growth-promoting bacteria to be used as inoculants in beneficial agricultural technologies. Microbiological Research. 2019;**219**:12-25

[3] Etesami H. Enhanced phosphorus fertilizer use efficiency with microorganisms. In: Meena RS, editor. Nutrient Dynamics for Sustainable Crop Production. Singapore: Springer Singapore; 2019. pp. 215-245

[4] Rouphael Y, Colla G. Toward a sustainable agriculture through plant biostimulants: From experimental data to practical applications. Agronomy;**10**:1461. Epub ahead of print 2020. DOI: 10.3390/agronomy10101461

[5] Mehmood T, Gaurav GK, Cheng L, et al. A review on plant-microbial interactions, functions, mechanisms and emerging trends in bioretention system to improve multi-contaminated stormwater treatment. Journal of Environmental Management. 2021;**294**:113108

[6] Bhatt K, Suyal DC, Kumar S, et al. New insights into engineered plant-microbe interactions for pesticide removal. Chemosphere. 2022;**309**:136635

[7] Doornbos RF, Geraats BPJ, Kuramae EE, et al. Effects of Jasmonic acid, ethylene, and salicylic acid signaling on the rhizosphere bacterial community of *Arabidopsis thaliana*. /395 MPMI. 2011;**24**:395-407

[8] Basu S, Rabara RC, Negi S, et al. Engineering PGPMOs through gene editing and systems biology: A solution for phytoremediation? Trends in Biotechnology. 2018;**36**:499-510

[9] Dhawi F. The role of plant growth-promoting microorganisms (PGPMs) and their feasibility in hydroponics and vertical farming. Metabolites;**13**:247. Epub ahead of print 1 February 2023. DOI: 10.3390/metabo13020247

[10] Cao Y, Wu X, Zhukova A, et al. Arbuscular mycorrhizal fungi (AMF) species and abundance exhibit different effects on saline-alkaline tolerance in *Leymus chinensis*. Journal of Plant Interactions. 2020;**15**:266-279

[11] Kuila D, Ghosh S. Aspects, problems and utilization of Arbuscular Mycorrhizal (AM) application as bio-fertilizer in sustainable agriculture. Current Research in Microbial Sciences. 2022;**3**:100107

[12] Kundu P, Nehra A, Gill R, et al. Genomic collation revealed the significance of *Piriformospora indica (Serendipita indica)* PiEF-hand protein in vesicle trafficking and fungal hyphal growth. South African Journal of Botany. 2022;**150**:548-564

[13] Singhal U, Prasad R, Varma A. *Piriformospora indica (Serendipita indica)*: The novel Symbiont. In: Mycorrhiza - Function, Diversity, State of the Art. Cham: Springer International Publishing; 2017. pp. 349-364

[14] Verma S, Varma A, Rexer K-H, et al. *Piriformospora indica*, gen. Et sp.

nov., a new root-colonizing fungus. Mycologia. 1998;**90**:896-903

[15] Bayubaskara MF, Ohme-Takagi M, Chan M-T. Optimal culture conditions of and Piriformospora indica for volatile compound release to promote effective plant growth. Plant Biotechnology. 2023;**40**:22.1221a

[16] Osman M, Stigloher C, Mueller MJ, et al. An improved growth medium for enhanced inoculum production of the plant growth-promoting fungus *Serendipita indica*. Plant Methods. 2020;**16**:39

[17] Saleem S, Sekara A, Pokluda R. Serendipita indica-a review from agricultural point of view. Plants;**10**:3417. Epub ahead of print 2022. DOI: 10.3390/xxxxx

[18] Rokni N, Shams Alizadeh H, Bazgir E, et al. The tripartite consortium of *Serendipita indica*, *Trichoderma simmonsii*, and bell pepper (*Capsicum annum*). Biological Control. 2021;**158**:104608

[19] Saleem S, Ragasova LN, Tekielska D, et al. *Serendipita indica* as a plant growth promoter and biocontrol agent against black rot disease in cabbage grown in a Phytotron. Agriculture. 2023;**13**:2048

[20] Fu W-L, Wu W-J, Xiao Z-Y, et al. *Serendipita indica*: A promising biostimulant for improving growth, nutrient uptake, and sugar accumulation in Camellia oleifera. Horticulturae. 2024;**10**:936

[21] Sehar S, Feng Q, Adil MF, et al. Tandem application of endophytic fungus *Serendipita indica* and phosphorus synergistically recuperate arsenic induced stress in rice. Frontiers in Plant Science;**13**. Epub ahead of print 6 September 2022. DOI: 10.3389/fpls.2022.982668

[22] Mosaddeghi MR, Hosseini F, Hajabbasi MA, et al. Epichloë Spp. and Serendipita Indica Endophytic Fungi: Functions in Plant-Soil Relations. Advances in Agronomy- Elsevier; 2021. pp. 59-113

[23] Charura N, Llamas E, De Quattro C, et al. Root cap cell corpse clearance limits microbial colonization in Arabidopsis thaliana. eLife;**13**:RP96266. Epub ahead of print 12 November 2024. DOI: 10.7554/eLife.96266

[24] Opitz MW, Díaz-Manzano FE, Ruiz-Ferrer V, et al. The other side of the coin: Systemic effects of *Serendipita indica* root colonization on development of sedentary plant–parasitic nematodes in Arabidopsis thaliana. Planta. 2024;**259**:121

[25] Boorboori MR, Zhang H-Y. The role of *Serendipita indica* (Piriformospora indica) in improving plant resistance to drought and salinity stresses. Biology (Basel). 2022;**11**:952

[26] Gill SS, Gill R, Trivedi DK, et al. *Piriformospora indica*: Potential and significance in plant stress tolerance. Frontiers in Microbiology. **7**:332. Epub ahead of print 22 March 2016. DOI: 10.3389/fmicb.2016.00332

[27] Tyagi J, Chaudhary P, Mishra A, et al. Role of endophytes in abiotic stress tolerance: With special emphasis on *Serendipita indica*. International Journal of Environmental Research. 2022;**16**:62

[28] Liu B, Liu X, Yu L, et al. *Serendipita indica* improve seed germination and seedling growth of *Lolium multiflorum* lam. Through amelioration of osmotic adjustment, nutrient accumulation and under salinity conditions. Plant, Soil and Environment. 2024;**70**:644-655

[29] Yang L, Zou Y-N, Tian Z-H, et al. Effects of beneficial endophytic fungal

inoculants on plant growth and nutrient absorption of trifoliate orange seedlings. Scientia Horticulturae. 2021;**277**:109815

[30] Nair DS, Manjula S. Induction of root endosymbiosis as a highly sustainable and efficient strategy for overproduction of the medicinally important diterpenoid lactone-andrographolide in *Andrographis paniculata* (Burm. F.) wall. Ex Nees. Industrial Crops and Products. 2020;**156**:112835

[31] Bouzouina M, Kouadria R, Lotmani B. Fungal endophytes alleviate salt stress in wheat in terms of growth, ion homeostasis and osmoregulation. Journal of Applied Microbiology. 2021;**130**:913-925

[32] Abdolmaleki AK, Pirdashti H, Yaghoubian Y, et al. Endophytic fungi improve growth and yield of wheat (*Triticum aestivum* L.) under limited light conditions. Gesunde Pflanzen. 2023;**75**:1517-1529

[33] Cao M-A, Liu R-C, Xiao Z-Y, et al. Symbiotic fungi alter the acquisition of phosphorus in Camellia oleifera through regulating root architecture, plant phosphate transporter gene expressions and soil phosphatase activities. Journal of Fungi. 2022;**8**:800

[34] Kundu A, Vadassery J. Molecular mechanisms of *Piriformospora indica* mediated growth promotion in plants. Plant Signaling and Behavior. **17**:2096785. Epub ahead of print 31 December 2022. DOI: 10.1080/15592324.2022.2096785

[35] Rani M, Raj S, Dayaman V, et al. Functional characterization of a hexose transporter from root Endophyte *Piriformospora indica*. Frontiers in Microbiology. 7:01083. Epub ahead of print 22 July 2016. DOI: 10.3389/fmicb.2016.01083

[36] Sherameti I, Shahollari B, Venus Y, et al. The endophytic fungus *Piriformospora indica* stimulates the expression of nitrate Reductase and the starch-degrading enzyme Glucan-water Dikinase in tobacco and Arabidopsis roots through a Homeodomain transcription factor that binds to a conserved motif in their promoters. Journal of Biological Chemistry. 2005;**280**:26241-26247

[37] Cruz C, Fegghi Z, Martins-Loução MA, et al. Plant Nitrogen Use Efficiency May Be Improved through Symbiosis with Piriformospora Indica. Sebacinales and Their Biotechnological Applications. Berlin, Heidelberg: Springer; 2013. pp. 285-293

[38] Prasad D, Verma N, Bakshi M, et al. Functional characterization of a magnesium transporter of root endophytic fungus *Piriformospora indica*. Frontiers in Microbiology. **9**:03231. Epub ahead of print 9 January 2019. DOI: 10.3389/fmicb.2018.03231

[39] Narayan OP, Verma N, Jogawat A, et al. Sulfur transfer from the endophytic fungus *Serendipita indica* improves maize growth and requires the sulfate transporter SiSulT. Plant Cell. 2021;**33**:1268-1285

[40] Zuccaro A, Lahrmann U, Güldener U, et al. Endophytic life strategies decoded by genome and transcriptome analyses of the mutualistic root symbiont *Piriformospora indica*. PLoS Pathogens. 2011;**7**:e1002290

[41] Conchillo LB, Haro R, Benito B. K+ nutrition exchange in the Serendipita-Arabidopsis symbiosis: Study of the fungal K+ transporters involved. Frontiers in Ecology and Evolution. **9**:789371. Epub ahead of print 16 December 2021. DOI: 10.3389/fevo.2021.789371

[42] González Ortega-Villaizán A, King E, Patel MK, et al. The endophytic fungus *Serendipita indica* affects auxin distribution in *Arabidopsis thaliana* roots through alteration of auxin transport and conjugation to promote plant growth. Plant, Cell & Environment. 2024;**47**:3899-3919

[43] Xu L, Wu C, Oelmüller R, et al. Role of phytohormones in *Piriformospora indica*-induced growth promotion and stress tolerance in plants: More questions than answers. Frontiers in Microbiology. **9**:01646. Epub ahead of print 31 July 2018. DOI: 10.3389/fmicb.2018.01646

[44] Pan R, Xu L, Wei Q, et al. *Piriformospora indica* promotes early flowering in Arabidopsis through regulation of the photoperiod and gibberellin pathways. PLoS One. 2017;**12**:e0189791

[45] Vahabi K, Sherameti I, Bakshi M, et al. The interaction of Arabidopsis with *Piriformospora indica* shifts from initial transient stress induced by fungus-released chemical mediators to a mutualistic interaction after physical contact of the two symbionts. BMC Plant Biology. 2015;**15**:58

[46] Godoy F, Olivos-Hernández K, Stange C, et al. Abiotic stress in crop species: Improving tolerance by applying plant metabolites. Plants. 2021;**10**:186

[47] Yaghoubian Y, Siadat SA, Moradi Telavat MR, et al. Bio-removal of cadmium from aqueous solutions by filamentous fungi: Trichoderma spp. and *Piriformospora indica*. Environmental Science and Pollution Research. 2019;**26**:7863-7872

[48] Ghabooli M, Kaboosi E. Alleviation of the adverse effects of drought stress using a desert adapted endophytic fungus and glucose in tomato. Rhizosphere. 2022;**21**:100481

[49] Wang Y, Cao J-L, Hashem A, et al. *Serendipita indica* mitigates drought-triggered oxidative burst in trifoliate orange by stimulating antioxidant defense systems. Frontiers in Plant Science. **14**:1247342. Epub ahead of print 4 October 2023. DOI: 10.3389/fpls.2023.1247342

[50] Hosseini F, Mosaddeghi MR, Dexter AR. Effect of the fungus *Piriformospora indica* on physiological characteristics and root morphology of wheat under combined drought and mechanical stresses. Plant Physiology and Biochemistry. 2017;**118**:107-120

[51] Saddique MAB, Ali Z, Khan AS, et al. Inoculation with the endophyte *Piriformospora indica* significantly affects mechanisms involved in osmotic stress in rice. Rice. 2018;**11**:34

[52] Xu L, Wang A, Wang J, et al. *Piriformospora indica* confers drought tolerance on Zea mays L. through enhancement of antioxidant activity and expression of drought-related genes. Crop Journal. 2017;**5**:251-258

[53] Zhang W, Wang J, Xu L, et al. Drought stress responses in maize are diminished by *Piriformospora indica*. Plant Signaling and Behavior. 2018;**13**:e1414121

[54] Hosseini F, Mosaddeghi MR, Dexter AR, et al. Maize water status and physiological traits as affected by root endophytic fungus *Piriformospora indica* under combined drought and mechanical stresses. Planta. 2018;**247**:1229-1245

[55] Tsai H-J, Shao K-H, Chan M-T, et al. *Piriformospora indica* symbiosis improves water stress tolerance of rice through regulating stomata behavior and ROS

scavenging systems. Plant Signaling & Behavior. 2020;**15**:1722447

[56] Mohsenifard E, Ghabooli M, Mehri N, et al. Regulation of miR159 and miR396 mediated by Piriformospora indica confer drought tolerance in rice. Journal of Plant Molecular Breeding. 2017;**5**:10-18

[57] Porcel R, Aroca R, Ruiz-Lozano JM. Salinity stress alleviation using arbuscular mycorrhizal fungi. A review. Agronomy for Sustainable Development. 2012;**32**:181-200

[58] Ghorbani A, Razavi SM, Ghasemi Omran VO, et al. *Piriformospora indica* inoculation alleviates the adverse effect of NaCl stress on growth, gas exchange and chlorophyll fluorescence in tomato (*Solanum lycopersicum* L.). Plant Biology. 2018;**20**:729-736

[59] Abdelaziz ME, Abdelsattar M, Abdeldaym EA, et al. *Piriformospora indica* alters Na+/K+ homeostasis, antioxidant enzymes and LeNHX1 expression of greenhouse tomato grown under salt stress. Scientia Horticulturae. 2019;**256**:108532

[60] Abdel Latef AAH, Chaoxing H. Effect of arbuscular mycorrhizal fungi on growth, mineral nutrition, antioxidant enzymes activity and fruit yield of tomato grown under salinity stress. Scientia Horticulturae. 2011;**127**:228-233

[61] Sharma P, Kharkwal AC, Abdin MZ, et al. *Piriformospora indica*-mediated salinity tolerance in Aloe vera plantlets. Symbiosis. 2017;**72**:103-115

[62] Sepehri M, Ghaffari MR, Khayam Nekoui M, et al. Root endophytic fungus *Serendipita indica* modulates barley leaf blade proteome by increasing the abundance of photosynthetic proteins in response to salinity. Journal of Applied Microbiology. 2021;**131**:1870-1889

[63] de Zélicourt A, Synek L, Saad MM, et al. Ethylene induced plant stress tolerance by Enterobacter sp. SA187 is mediated by 2-keto-4-methylthiobutyric acid production. PLoS Genetics. 2018;**14**:e1007273

[64] Jogawat A, Vadassery J, Verma N, et al. PiHOG1, a stress regulator MAP kinase from the root endophyte fungus *Piriformospora indica*, confers salinity stress tolerance in rice plants. Scientific Reports. 2016;**6**:36765

[65] Lanza M, Haro R, Conchillo LB, et al. The endophyte *Serendipita indica* reduces the sodium content of Arabidopsis plants exposed to salt stress: Fungal ENA ATPases are expressed and regulated at high pH and during plant co-cultivation in salinity. Environmental Microbiology. 2019;**21**:3364-3378

[66] Zarea MJ, Hajinia S, Karimi N, et al. Effect of *Piriformospora indica* and Azospirillum strains from saline or non-saline soil on mitigation of the effects of NaCl. Soil Biology and Biochemistry. 2012;**45**:139-146

[67] Yang M, Fang Y, Sun D, et al. Efficiency of two nitrification inhibitors (dicyandiamide and 3, 4-dimethypyrazole phosphate) on soil nitrogen transformations and plant productivity: A meta-analysis. Scientific Reports. **6**:22075. Epub ahead of print 23 February 2016. DOI: 10.1038/srep22075

[68] Varma A, Bakshi M, Lou B, et al. *Piriformospora indica*: A novel plant growth-promoting Mycorrhizal fungus. Agricultural Research. 2012;**1**:117-131

[69] Peskan-Berghöfer T, Vilches-Barro A, Müller TM, et al. Sustained exposure to abscisic acid enhances the colonization

potential of the mutualist fungus *Piriformospora indica* on *Arabidopsis thaliana* roots. New Phytologist. 2015;**208**:873-886

[70] Hassani D, Khalid M, Huang D, et al. Morphophysiological and molecular evidence supporting the augmentative role of in mitigation of salinity in L. Acta Biochim Biophys Sin (Shanghai). 2019;**51**:301-312

[71] Khalid M, Hassani D, Liao J, et al. An endosymbiont *Piriformospora indica* reduces adverse effects of salinity by regulating cation transporter genes, phytohormones, and antioxidants in *Brassica campestris ssp. Chinensis*. Environmental and Experimental Botany. 2018;**153**:89-99

[72] Ghorbani A, Razavi SM, Omran VOG, et al. Piriformospora indica alleviates salinity by boosting redox poise and antioxidative potential of tomato. Russian Journal of Plant Physiology. 2018;**65**:898-907

[73] Manghwar H, Hussain A, Alam I, et al. Waterlogging stress in plants: Unraveling the mechanisms and impacts on growth, development, and productivity. Environmental and Experimental Botany. 2024;**224**:105824

[74] Liang S-M, Hashem A, Abd-Allah EF, et al. Root-associated symbiotic fungi enhance waterlogging tolerance of peach seedlings by increasing flavonoids and activities and gene expression of antioxidant enzymes. Chemical and Biological Technologies in Agriculture. 2023;**10**:124

[75] Li L, Feng Y, Qi F, et al. Research Progress of *Piriformospora indica* in improving plant growth and stress resistance to plant. Journal of Fungi. 2023;**9**:965

[76] Wan Y-X, Kapoor R, da Silva FSB, et al. Elucidating the mechanism regarding enhanced tolerance in plants to abiotic stress by *Serendipita indica*. Plant Growth Regulation. 2024;**103**:271-281

[77] Liang S, Yingning ZO, Bo SH, Qiangsheng WU. Arbuscular mycorrhizal fungi and endophytic fungi differentially modulate polyamines or proline of peach in response to soil flooding. Pedosphere. 2024;**34**(2):460-472

[78] Jiang W, Pan R, Wu C, et al. *Piriformospora indica* enhances freezing tolerance and post-thaw recovery in *Arabidopsis* by stimulating the expression of *CBF* genes. Plant Signaling and Behavior. 2020;**15**:1745472

[79] Adhikari L, Baral R, Paudel D, et al. Cold stress in plants: Strategies to improve cold tolerance in forage species. Plant Stress. 2022;**4**:100081

[80] Wang Y, Wang J, Sarwar R, et al. Research progress on the physiological response and molecular mechanism of cold response in plants. Frontiers in Plant Science;**15**:271-281. Epub ahead of print 30 January 2024. DOI: 10.3389/fpls.2024.1334913

[81] Shen J, Chen Y. Serendipita indica: A biostimulant enhancing low-temperature tolerance and active constituent levels in *Polygonum cuspidatum*. Agriculture. 2023;**14**:7

[82] Jiang W, Pan R, Buitrago S, et al. Transcriptome analysis of Arabidopsis reveals freezing-tolerance related genes induced by root endophytic fungus *Piriformospora indica*. Physiology and Molecular Biology of Plants. 2021;**27**:189-201

[83] Feng D, Wang R, Sun X, et al. Heavy metal stress in plants: Ways to alleviate

with exogenous substances. Science of the Total Environment. 2023;**897**:165397

[84] Liu S, Yang R, Pan Y, et al. Nitric oxide contributes to minerals absorption, proton pumps and hormone equilibrium under cadmium excess in *Trifolium repens* L. plants. Ecotoxicology and Environmental Safety. 2015;**119**:35-46

[85] AbdElgawad H, Zinta G, Hamed BA, et al. Maize roots and shoots show distinct profiles of oxidative stress and antioxidant defense under heavy metal toxicity. Environmental Pollution. 2020;**258**:113705

[86] Sharaya R, Deepa NA, et al. *Piriformospora indica (Serendipita indica)*: Potential tool for alleviation of heavy metal toxicity in plants. In: Biostimulants in Alleviation of Metal Toxicity in Plants. NJ: Elsevier; 2023. pp. 401-422

[87] Li L, Zhu P, Wang X, et al. Phytoremediation effect of *Medicago sativa* colonized by Piriformospora indica in the phenanthrene and cadmium co-contaminated soil. BMC Biotechnology. 2020;**20**:20

[88] Dhiman K, Sharma D, Kumari R, et al. Biofortification of crops using microbes – A promising sustainable agriculture strategy. Journal of Plant Nutrition. 2023;**46**:2912-2935

[89] Oelmüller R, Sherameti I, Tripathi S, et al. *Piriformospora indica*, a cultivable root endophyte with multiple biotechnological applications. Symbiosis. 2009;**49**:1-17

[90] Li Q-S, Xie Y-C, Rahman MM, et al. Arbuscular mycorrhizal fungi and endophytic fungi activate leaf antioxidant defense system of lane late navel orange. Journal of Fungi. 2022;**8**:282

[91] Khalvandi M, Amerian M, Pirdashti H, et al. Essential oil of peppermint in symbiotic relationship with *Piriformospora indica* and methyl jasmonate application under saline condition. Industrial Crops and Products. 2019;**127**:195-202

[92] Serfling A, Wirsel SGR, Lind V, et al. Performance of the biocontrol fungus *Piriformospora indica* on wheat under greenhouse and field conditions. Phytopathology. 2007;**97**:523-531

[93] Smriti S, Ajit V. From *Piriformospora indica* to rootonic: A review. African Journal of Microbiology Research. 2014;**8**:2984-2992

Chapter 7

Perspective Chapter: Unlocking Root Architecture Plasticity – Enhancing Crop Productivity and Resilience to Abiotic Stresses

Ilva Licaj, Tahreem Ahmed and Mariapina Rocco

Abstract

The escalating global population has heightened the demand for food and feed, making the effects of abiotic stresses on crop yield and quality a significant challenge for global food security. Environmental stressors, such as drought, salinity, heat, disturb plant metabolic functions, redirecting energy from growth processes toward stress adaptation, which ultimately reduces plant productivity. The development of crop varieties resilient to these abiotic stresses is imperative to address the urgent need for sustainable food production. Recent advancements highlight the potential of crops with improved root systems, supported by phytohormone regulation, to enhance water and nutrient uptake, sequester more carbon, and demonstrate greater resilience to climate change. In turn, these increase the potential for crops to stabilize yields under fluctuating environmental conditions. This chapter examines the physiological, molecular, and genetic mechanisms that govern root system architecture (RSA), focusing on the regulatory pathways that influence root growth, branching, and overall adaptability to stress. Furthermore, the chapter discusses the application of advanced phenotyping that has revolutionized our ability to study root traits linked to stress resilience. The integration of this technology with molecular genetics and breeding strategies is accelerating the development of climate-resilient crops.

Keywords: abiotic stresses, drought stress, root system architecture (RSA), plant phenotyping, reactive oxygen species (ROS), genomic

1. Introduction

The accelerating pace of global climate change is intensifying the frequency, duration, and severity of abiotic stresses, including drought, heat waves, soil degradation, and nutrient depletion. These multifaceted stressors are increasingly acting in combination, posing serious threats to the stability and productivity of agricultural systems worldwide [1, 2]. Drought, in particular, stands out as one of the most pervasive and economically damaging abiotic challenges. According to FAO estimates, between 2005 and 2015, drought-induced agricultural losses exceeded US$29 billion in developing regions alone [3]. The risk is further amplified by the co-occurrence of drought with other stressors, such as elevated temperatures, salinity, and limited nutrient availability, which together compound physiological stress and reduce crop yields [1, 4].

At the physiological level, drought affects plants primarily through reduced soil water availability, leading to a cascade of negative effects. Limited water supply disrupts nutrient diffusion and mineralization, particularly for nitrogen (N) and phosphorus (P)—nutrients essential for photosynthesis, biomass accumulation, and reproductive development [5–7]. Drought also imposes hydraulic limitations, impairs cell elongation, reduces stomatal conductance, and initiates oxidative stress *via* the accumulation of reactive oxygen species (ROS) that can damage membranes, proteins, and DNA [8, 9]. In response, plants activate complex signaling networks involving abscisic acid (ABA), Ca^{2+} signaling, and ROS detoxification, leading to adaptive responses that include morphological, physiological, and metabolic adjustments [2, 10].

While aboveground traits such as canopy architecture and transpiration efficiency contribute to drought tolerance, root traits are increasingly recognized as critical determinants of plant resilience under water-limited conditions [11]. Roots act as both sensors and responders to drought stress, with their ability to dynamically adjust structure and function providing the foundation for adaptive responses. Central to this capacity is the root system architecture (RSA), a term encompassing root depth, length, growth angle, lateral branching, diameter, and the development of root hairs and nodal roots. Under drought conditions, RSA typically undergoes plastic changes, for example, the promotion of deeper, steeper-angled primary roots and the suppression of shallow lateral roots, thereby enhancing the plant's ability to access water from deeper soil layers [12]. These architectural shifts are often driven by alterations in the gravitropic set-point angle (GSA) and are genetically regulated. For instance, the DRO1 gene in rice modifies root angle to facilitate deeper rooting and has been successfully leveraged in breeding programs to enhance drought tolerance [13, 14]. Similar efforts are underway in sorghum, maize, and wheat. Other root traits such as root hair length and density contribute by expanding the root-soil interface, lowering hydraulic resistance, and improving nutrient uptake under drought conditions [15]. Root anatomical features, such as xylem vessel diameter, aerenchyma formation, and cortical cell size, also influence water uptake efficiency and drought resilience, although these remain less commonly targeted in breeding efforts [16].

Historically underrepresented in trait-based breeding, roots are now at the forefront of efforts to develop climate-resilient crops. Advances in phenotyping technologies, such as X-ray computed tomography, ground-penetrating radar, and shovelomics, coupled with molecular tools like Quantitative Trait Loci (QTL) mapping and CRISPR/Cas9 gene editing, have enabled detailed dissection of RSA traits [17]. These tools are helping to identify key loci and gene networks associated with drought-adaptive root traits, facilitating their integration into genomic selection and marker-assisted breeding.

However, significant challenges remain in translating laboratory findings to field conditions. Root traits often show genotype-by-environment (G × E) interactions, and their expression is influenced by complex soil-plant-atmosphere dynamics. Moreover, trade-offs between drought resilience and other agronomic traits, such as nutrient uptake or mechanical stability, must be carefully managed to ensure overall crop performance.

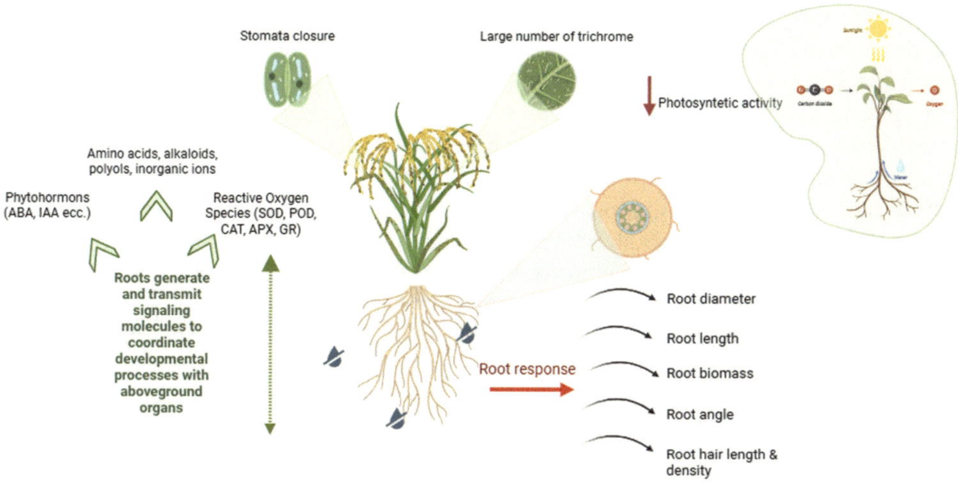

Figure 1.
A diagrammatic illustration of physiological, anatomical, and morphological traits of the root system architecture influencing root development.

In light of escalating climate variability, understanding and harnessing the functional diversity of root system architecture offers a promising path toward drought-resilient agriculture. RSA traits not only determine a plant's ability to capture water and nutrients but also interact with microbial communities, influence carbon sequestration, and affect long-term soil health. This chapter explores how RSA contributes to drought adaptation from a multidisciplinary perspective, encompassing genetics, physiology, ecology, and agronomy (**Figure 1**). By bridging basic research and applied breeding, we aim to highlight RSA's pivotal role in securing future crop productivity under increasingly challenging environmental conditions.

2. Morpho-functional plasticity of roots in response to drought stress

Root plasticity, defined as the capacity of a plant's root system to undergo morphological and functional adjustments in response to environmental stimuli, is increasingly recognized as a key trait underlying drought adaptation [18, 19]. Soil type and its physical characteristics play a fundamental role in shaping root responses to drought. Attributes such as texture, bulk density, compaction, porosity, water-holding capacity, and mechanical impedance significantly influence how roots grow, orient, and function in a given environment [20, 21]. These factors determine both the availability of water and the physical space through which roots can proliferate, thereby guiding the spatial configuration of the root system.

In coarse-textured soils, such as sandy or sandy-loam types—where water drains rapidly and retention is low—early development of deep rooting systems is advantageous. These roots have long been viewed as a desirable trait for drought tolerance, particularly in crops grown in regions with deeper soil moisture reserves [22], since they can access residual moisture in the subsoil, improving both yield and water-use efficiency under drought conditions. However, emerging research highlights the situational benefits of shallow rooting. In Arabidopsis, shallow-rooted genotypes captured transient surface moisture more efficiently under certain drought conditions, mirroring strategies observed in desert-adapted species [11]. This challenges the "deeper is better" paradigm, suggesting that rooting depth must be optimized to match specific environmental contexts, including rainfall frequency, soil depth, and crop system design [23, 24].

By contrast, in fine-textured soils, such as clay or silty soils, moisture is more likely to be retained in the upper soil horizons. As a result, plants tend to allocate more root growth near the surface to exploit these transient water resources [21]. In compacted soils or those with mechanical barriers such as hardpans, vertical root penetration is often restricted. Under these conditions, plants exhibit adaptive plasticity by enhancing lateral root proliferation near the surface, which can partially compensate for limited deep-water access [25, 26]. This shift is not just a structural adaptation but also reflects a strategic trade-off to maintain water and nutrient uptake under mechanical stress.

Moreover, plant life history and growth form further modulate root trait plasticity. Annual species typically display higher root plasticity than perennials, enabling them to adjust rapidly to fluctuating soil moisture regimes during short life cycles [27, 28]. Among root types, fibrous-rooted species, such as cereals and grasses, exhibit greater responsiveness to soil moisture gradients due to their highly branched and diffuse root networks. In contrast, tap-rooted species, such as many legumes and dicots, tend to exhibit less plasticity, though their deep-penetrating primary roots inherently allow access to deeper water reserves [29].

To develop drought-resilient crops, a deeper understanding of the spatial and temporal dynamics of root traits is essential. RSA, the three-dimensional configuration of the root system in the soil, includes traits such as root growth angle, depth, length, diameter, root hair density, and root-to-shoot biomass ratio. Each of these traits contributes to a plant's ability to access water and nutrients under limited supply conditions. Root growth angle is a particularly influential trait, governing the depth at which roots explore the soil profile. Controlled by gravitropic responses and shaped by both genetic and environmental cues, steeper root angles lead to deeper rooting, which has been positively correlated with drought tolerance and improved grain yield in crops such as rice, wheat, and maize [30, 31]. Root-to-shoot biomass allocation is often altered under drought stress. A higher root-to-shoot ratio reflects a drought avoidance strategy, prioritizing below-ground growth to enhance resource capture. However, this response is highly context-dependent. While moderate drought can stimulate root growth, severe stress may reduce root biomass due to carbon limitations or shift the strategy toward shoot regrowth post-stress [32, 33]. Root length and diameter also contribute to drought resilience. Longer roots allow deeper soil exploration, expanding the zone of water access, but at a higher metabolic cost. Finer roots, with a higher surface area-to-volume ratio, are more efficient for rapid water uptake, while thicker roots often support structural stability and enhanced branching for sustained uptake [34–37].

Although root hairs represent only a minor portion of total root biomass, they play a disproportionately large role in water and nutrient acquisition, especially under drought. Their density and elongation frequently increase in response to water stress, expanding the effective absorptive surface area of the root system and improving hydraulic conductivity [38, 39]. Experimental studies have shown that barley mutants lacking root hairs exhibit reduced water uptake despite having highly branched root systems, emphasizing the importance of root hairs in drought responses [15]. Root hairs also contribute significantly to rhizosheath formation, the zone of soil closely adhered to the root surface. This structure retains moisture and supports beneficial microbial colonization, which can enhance drought tolerance through improved nutrient cycling, hormone production, and stress mitigation [40, 41].

3. Adaptations of plants to water-limited environments: Integrating physiological, biochemical, and nutritional responses

Under water-limited conditions, plants modulate root development dynamically. Plants inhabiting drought-prone environments deploy a sophisticated array of biochemical and physiological strategies to maintain cellular turgor and protect vital functions under water scarcity. Beyond structural adaptation, root-to-shoot communication plays a central role in plant drought responses [42]. Roots perceive changes in soil moisture and transmit hormonal signals such as abscisic acid (ABA), cytokinins, and peptides to regulate stomatal conductance, leaf growth, and overall water balance [43, 44]. Hydraulic signals, triggered by tension changes in the xylem, also provide rapid feedback to aboveground tissues during soil drying. The integration of these signaling pathways enables coordinated drought responses regarding root development, water uptake, and stress tolerance at multiple organizational levels, across the entire plant.

In response to drought-induced osmotic stress, plants accumulate compatible solutes such as proline, which stabilize proteins and membranes, protect enzyme function, and improve intracellular water retention [44, 45]. Elevated proline levels under drought have been documented in numerous crop species, highlighting its universal role in osmoprotection [46]. Concomitantly, drought induces oxidative stress *via* excessive reactive oxygen species (ROS) production, necessitating robust antioxidant defenses. Enzymatic antioxidants like superoxide dismutase (SOD) and peroxidases convert harmful ROS into less reactive molecules, while nonenzymatic antioxidants, including ascorbate, glutathione, and flavonoids, further scavenge free radicals [47]. Secondary metabolites such as phenolics enhance drought tolerance by reinforcing cell walls through cross-linking, reducing cellular water loss, and providing mechanical strength (**Figure 2**) [48, 49].

At the core of drought adaptation lies a dynamic hormonal interplay, predominantly involving ABA, auxin, and cytokinin. This hormonal crosstalk intricately modulates RSA and physiological responses, allowing plants to optimize water acquisition and growth despite fluctuating soil moisture [50, 51]. The balance between antagonistic and synergistic hormone interactions is critical for fine-tuning root growth direction, branching patterns, and functional specialization in response to drought [52].

ABA is widely recognized as the principal hormone orchestrating drought responses. Its biosynthesis is rapidly induced by dehydration signals primarily sensed in the leaves, which communicates with roots to trigger adaptive responses [53]. A hallmark of ABA function is xerobranching, suppression of lateral root formation in dry soil zones, thereby directing root proliferation toward wetter regions [40]. In maize and other crops, ABA modulates biomass allocation by promoting root elongation at the expense of shoot growth, thus enhancing soil water foraging capacity [54]. Beyond growth regulation, ABA signaling intersects with calcium-mediated pathways to facilitate hydrotropism—the directional bending of roots toward moisture gradients. Components such as MIZUKUSSEI1 (MIZ1) and SnRK2 kinases mediate asymmetric cytosolic Ca^{2+} fluxes in root elongation zones, translating soil water heterogeneity into targeted root growth responses (**Figures 3** and **4**) [55, 56]. These calcium spikes also activate calcium-dependent protein kinases (CDPKs) that

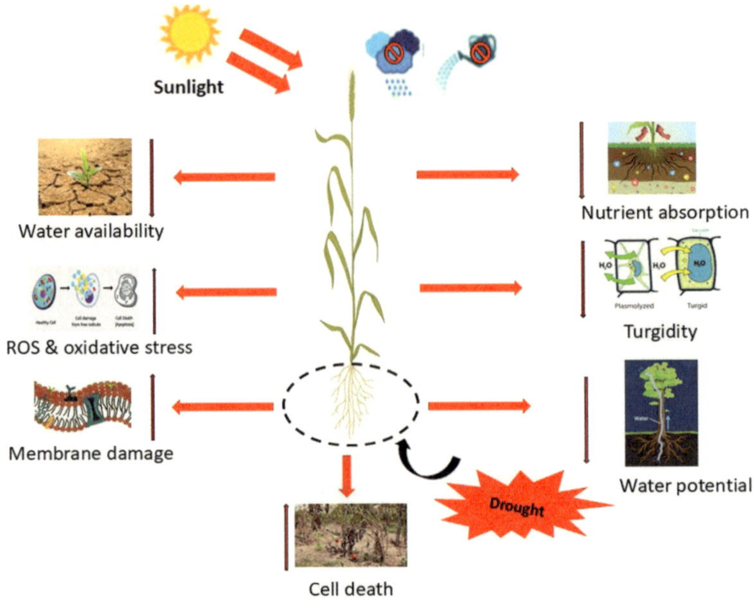

Figure 2.
Illustration of physiological and biochemical disruptions in plants under drought stress.

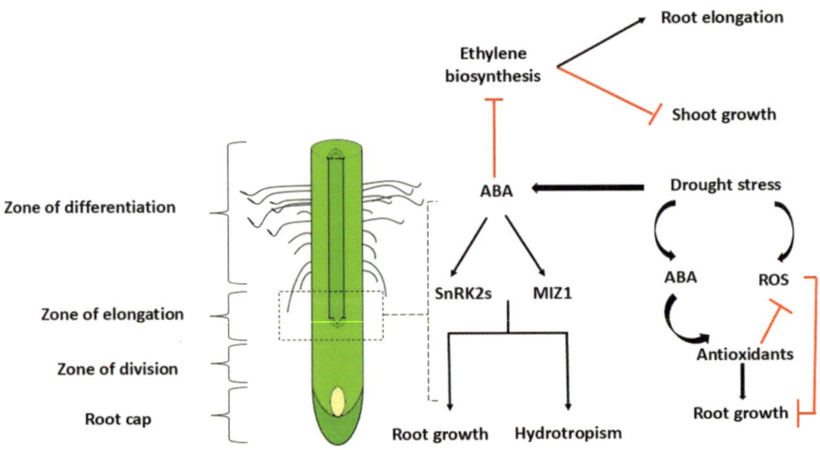

Figure 3.
Schematic representation of the regulatory network of hormonal signaling and oxidative stress influencing root growth and adaptation under drought stress conditions. Black arrows (→) indicate activation of a downstream process, while red blunt arrows (—┤) represent negative regulation. The root tip is anatomically divided into three zones: the differentiation zone, the elongation zone, the division zone, and the root cap. Drought stress induces ABA accumulation, which activates SnRK2s and MIZ1 to promote root growth and hydrotropism. ABA simultaneously inhibits ethylene biosynthesis, thus promotin-g ethylene-mediated suppression of root elongation while limiting shoot growth. ROS elevated under drought, triggering antioxidant responses that modulate root growth.

regulate ion channels, stomatal closure, and the expression of drought-responsive genes, integrating cellular and systemic drought signaling [57, 58]. Recent discoveries reveal that ABA controls root hydraulic properties at a molecular level by modulating microRNAs, such as miR165/166, which repress HD-ZIP III transcription factors,

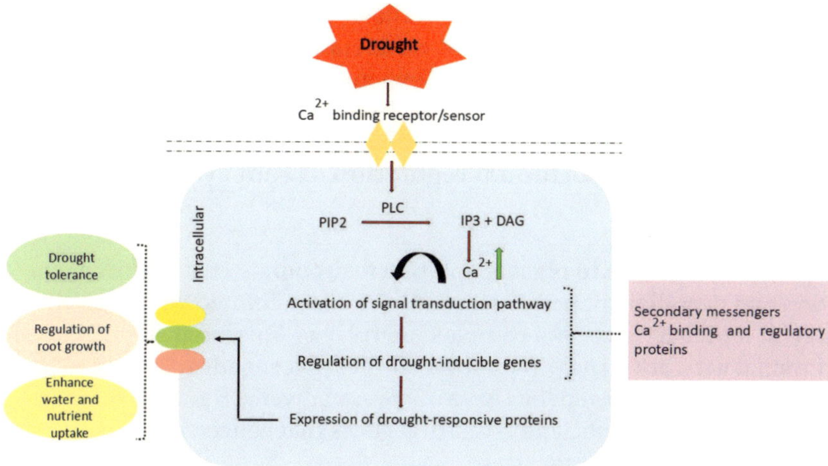

Figure 4.
Schematic illustration of Ca^{+2}-mediated signal transduction pathway activated under drought stress. Drought stress activates Ca^{+2}-binding receptors/sensors located at the plasma membrane to initiate the drought-responsive mechanism. This Ca^{+2}-mediated receptor activation leads to the activation of phospholipase C (PLC), which hydrolyzes phosphatidylinositol 4,5-bisphosphate (PIP2) into Inositol triphosphate (IP3) and Diacylglycerol (DAG). IP3 supports the release of Ca^{+2} from intracellular stores, thus increasing cytosolic Ca^{+2} levels. This elevated Ca^{+2} level then activates a signal transduction pathway, influenced by drought-sensitive secondary messengers, Ca^{+2}-binding proteins, and other regulatory proteins. This cascade modulates the transcription of drought-inducible genes, ultimately leading to the expression of drought-responsive proteins. These proteins are exported out of the cell via *channels that collectively regulate drought tolerance mechanisms, water status, nutrient uptake, and the regulation of root growth.*

promoting protoxylem differentiation and enhancing water conduction capacity under drought stress [59].

Auxin plays a pivotal role in shaping root growth patterns through mechanisms such as hydropatterning, hydrotropism, and xerotropism, allowing roots to spatially optimize water uptake [26, 60, 61]. It regulates gravitropic responses by controlling amyloplast sedimentation in columella cells, which influences root angle and lateral root formation [51]. Drought stress modifies auxin signaling pathways; for instance, SUMOylation of Auxin Response Factor 7 (ARF7) enhances its interaction with repressors such as IAA3, inhibiting lateral root development in dry zones and promoting growth in moist microsites [51]. This precise hormonal tuning ensures efficient root system remodeling aligned with soil water availability. Aquaporins (AQPs), integral membrane proteins regulated by auxin and other signals, adjust root hydraulic conductivity during drought, facilitating water flow through plasma membranes and vacuoles. Differential expression of plasma membrane intrinsic proteins (PIPs) and tonoplast intrinsic proteins (TIPs) under drought stress supports lateral root emergence and maintains root function [62, 63].

Moreover, mineral nutrition profoundly influences drought adaptation, often through modulating hormonal signaling and osmotic balance. Drought impairs nitrogen fixation in legumes, diminishing nitrogen supply and assimilating transport critical for seed filling [64–66]. Micronutrients, such as zinc, contribute to drought tolerance by promoting biosynthesis of growth regulators, osmolytes, and antioxidant enzymes [67, 68]. Phosphorus enhances root proliferation and hydraulic conductivity, facilitating nutrient uptake and improving yield under drought, as observed in wheat [69, 70]. Potassium supports osmotic regulation and membrane stability, key for maintaining proline accumulation and cellular integrity [71]. Magnesium aids in preserving photosynthetic efficiency and water status, while boron facilitates ROS detoxification and nutrient uptake, stabilizing membranes under drought [72, 73].

4. Transcriptional and hormonal regulation of root system architecture in drought resistance

RSA plays a pivotal role in plant adaptation to drought stress by influencing water acquisition and overall plant performance under water-limited conditions. RSA development is orchestrated by a complex interplay of hormonal signaling pathways, transcriptional networks, and environmental cues. Recent advances in functional genomics and quantitative genetics have uncovered diverse transcription factors (TFs), signaling components, and structural genes that collectively modulate RSA plasticity in response to drought stress.

WRKY TFs are pivotal in plant responses to various stresses, including drought [74–77]. They can function as activators or repressors and are involved in both abiotic and biotic stress responses [78]. For instance, the sorghum WRKY gene SbWRKY30 has been shown to confer drought tolerance in Arabidopsis and rice by affecting RSA, likely through enhanced reactive oxygen species (ROS) scavenging and the regulation of stress-responsive genes like SbRD29. In wheat, the WRKY TFs TaWRKY1 and TaWRKY33 have been implicated in drought and heat resistance [78, 79]. Overexpression of these genes in Arabidopsis enhanced germination rates, promoted root growth, and increased tolerance to dehydration and heat stress.

Auxin, a key phytohormone, interacts with gravity-responsive pathways to control root tip curvature, directing root growth either deep into soil strata or laterally across upper layers [80]. In rice, the DEEPER ROOTING 1 (DRO1) gene is a key determinant of root angle. As identified by Uga et al. [81], DRO1 influences auxin distribution at the root apex, enabling the formation of deeper roots that enhance water uptake and mitigate yield loss under drought [13]. Its homolog in maize, ZmDRO1, similarly promotes steeper root angles and improves drought resilience [82], indicating a conserved function across cereals [83].

Additional Quantitative Trait Loci (QTLs) in rice—DRO2, DRO3, and qSOR1—have also been associated with vertical or lateral root growth tendencies [84–87]. In *Arabidopsis*, the natural variation in EXOCYST70A3 regulates PIN4-dependent auxin localization, leading to adaptive root gravitropic responses and modulating rooting depth based on moisture availability [11]. Genetic association studies have further uncovered maize RSA regulators, such as ZmCIPK15 and ZmCIPK3, which modulate root angle and seminal root growth under drought [88, 89]. Historical selection during modern maize breeding has favored alleles that increase root depth, with ZmRSA3.1 and ZmRSA3.2 identified as auxin-related loci driving such changes [90].

Root TFs such as ZmPTF1, a bHLH protein, activate abscisic acid (ABA) biosynthesis by binding G-box motifs in NCED gene promoters, thereby enhancing root biomass and coordinating drought-responsive gene expression [46]. In maize, ZmbZIP4 enhances ABA accumulation and promotes root elongation, partly by repressing growth-inhibitory genes [91]. In rice, OsbZIP12 upregulates LEA3 and Rab16, contributing to improved drought tolerance, while OsbZIP52 represses these genes in an ABA-independent manner.

TFs such as OsERF48 and RRS1 (MYB-type) regulate lateral root initiation and primary root elongation by modulating auxin-responsive genes in rice [92, 93].

The *Arabidopsis* ARFs ARF7 and ARF19 act through LBD transcription factors, notably LBD16, to mediate hydropatterned lateral root growth toward water-rich zones [94, 95]. The R2R3-MYB family is highly responsive to drought in multiple crops. For example, AtMYB15 in *Arabidopsis* integrates ABA, ethylene, and jasmonic acid signaling, while MdMYB88 and MdMYB124 in apples enhance root-to-shoot ratios during prolonged drought stress [96].

HD-Zip TFs, particularly HAT1 and HTA3, function redundantly to regulate drought responses in *Arabidopsis*, with mutants showing heightened sensitivity [97]. The NF-Y complex (NF-YA/B/C) also contributes to RSA by regulating ABA-responsive and root-specific genes. Overexpression of NF-YA in *Arabidopsis* has been shown to enhance root growth and drought tolerance [98].

Beyond TFs, several key structural and enzymatic genes contribute to RSA plasticity and DR. ZmVPP1 and ZmTIP1 influence lateral root density and root hair length, respectively, while OsABA8ox2 knockout mutants accumulate ABA and auxin, increasing root biomass [99–102]. Manipulation of cytokinin metabolism *via* CKX1/3 also promotes root proliferation and DR [103]. Additional contributors include LRD (KNAT3 homolog) in wheat [104] and the *Arabidopsis* sugar transporter SWEET17, as well as SWEET11/12, which modulate root growth through ABA-dependent phosphorylation events [105, 106].

Anatomical traits, especially xylem development, are integral to improving water transport during drought. In rice, NAC-domain TFs such as OsNAC5, OsNAC9, and OsNAC10 promote root thickening and xylem expansion, enhancing drought tolerance [107, 108]. OsERF71 further promotes root diameter *via* cell wall remodeling and aerenchyma formation [109]. Conversely, XND1 inhibits xylem differentiation and aquaporin function, reducing root hydraulic conductivity under drought conditions [110].

Several genome-wide association studies (GWAS) and QTL analyses have identified key loci associated with drought-responsive RSA traits, notably in rice, wheat, and maize [111–115]. Traits such as root length and dry weight are consistently associated with DR, linked to QTLs like CRL1, PRL2/3, and SRL2/7/9 in maize and Qrdws.uwa-4AL/5AL in wheat [116, 117]. Whereas in rice, over 280 drought-regulated genes were revealed *via* GWAS, indicating divergence in water acquisition vs. conservation strategies [118]. Emerging evidence from under-studied species such as poplar, apple, and alfalfa further supports the universality of RSA-driven drought adaptation. In poplar, genes like PtabZIP1 and PagWOX11/12a enhance lateral root development and biomass under drought [119–121].

5. Root phenotyping under drought stress: Technologies, challenges, and prospects

Root phenotyping presents significant challenges for breeders aiming to improve drought stress tolerance in crops, primarily due to the hidden nature of roots and the complexity of their genetic control in field conditions [122]. Despite these difficulties, recent advances in imaging technologies and genomic tools have enhanced the interest in phenotyping root traits more accurately and efficiently, thereby accelerating our understanding of the genetic basis of root system architecture (RSA) and its adaptation to environmental stress (**Figure 5**).

Image-based phenotyping is a widely used, nondestructive approach that integrates optical imaging and computational analysis to characterize anatomical, physiological, and biochemical properties of roots [123–125]. Among the earliest

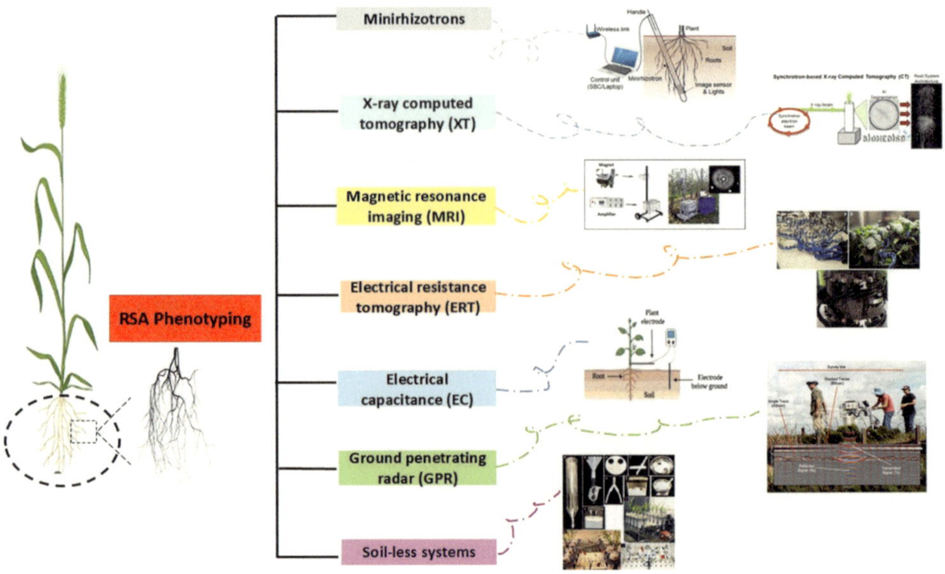

Figure 5.
Overview of advanced RSA phenotyping techniques for nondestructive analysis of root system architecture in crop plants.

techniques, minirhizotrons involve the installation of transparent tubes in the soil through which a camera captures images of root growth adjacent to the tube walls [126, 127]. However, this method may not reflect natural root development due to artificial soil-tube interfaces, potentially skewing growth observations [128].

More recent technologies include X-ray computed tomography (CT) and magnetic resonance imaging (MRI), which allow for high-resolution, three-dimensional visualization of root structures *in situ* without disturbing the soil matrix [129–132]. MRI quality, however, can be affected by soil water content and substrate type [133]. A notable application is the medium-throughput phenotyping pipeline established by Pflugfelder et al. [134], which used MRI to analyze RSA in 288 wheat seedlings, demonstrating the scalability of such techniques.

Electrical resistance tomography (ERT) and electrical capacitance (EC) offer indirect but insightful measurements of root-soil interactions. ERT monitors changes in soil resistivity as influenced by root activity and water movement [135, 136], while EC estimates root architecture based on the dielectric properties detected between the soil and plant stem [137]. EC-based methods have been successfully applied in field conditions across multiple crops, including maize, soybean, and wheat [138–140].

Ground-penetrating radar (GPR) is another powerful noninvasive tool that uses electromagnetic pulses to detect roots based on their dielectric contrast with the surrounding soil. GPR is particularly useful for identifying the number, diameter, and depth of coarse roots and has been effectively applied to crops such as cassava, citrus, and willow [141, 142]. These technologies support the assessment of RSA traits, such as root biomass and spatial distribution, without disturbing the soil profile [143].

In controlled environments, soil-less systems such as gels, aeroponics, and hydroponics enable precise phenotyping under standardized nutrient and moisture regimes. For example, transparent nail-board systems packed with glass beads have been used to image and measure root length, branching angles, and other RSA traits [144, 145].

These systems are ideal for early-stage root studies and functional characterization of root-specific genes.

Software tools have greatly enhanced the analysis of root images, allowing both manual and automated quantifications of RSA traits. Widely used programs include WinRHIZO, ImageJ, EZ-Rhizo, SmartRoot, RootNav, and RootTrace, each offering capabilities to analyze parameters such as root length, branching, diameter, and angle [146–148]. SmartRoot, in particular, facilitates analysis of root growth dynamics and angles, while DART, a Java-based manual tool, enables the construction of flexible datasets capturing individual root features [77].

6. Conclusion

Drought stress, triggered by reduced precipitation and elevated solar radiation, initiates a complex chain of physiological and biochemical events in plants. These include reduced water uptake, nutrient deficiencies, membrane destabilization, oxidative stress, and ultimately, tissue damage or programmed cell death. In this context, root system plasticity serves as a critical first line of defense, enabling plants to dynamically adjust root morphology, architecture, and function to maintain resource acquisition under water-limited conditions. At the core of this adaptive capacity is the root system architecture, a composite of traits such as root depth, angle, diameter, length, and root hair development. These traits are shaped by both genetic and environmental cues and are key determinants of water-use efficiency. Transcription factors such as WRKY, NAC, MYB, bZIP, HD-Zip, ARF, and NF-Y play central roles in regulating hormonal signaling pathways and developmental reprogramming under drought conditions. Alongside structural and anatomical genes, these regulators represent valuable targets for crop improvement. Recent advances in molecular biology and phenotyping technologies have begun to unravel the genetic bases of these RSA traits. High-throughput imaging, field-based phenotyping platforms, and quantitative trait mapping now allow for precise characterization and selection of drought-adaptive traits. To conclude, as climate change intensifies drought frequency and severity, the integration of genetic, physiological, and ecological knowledge is essential. Precision breeding, biotechnological interventions, and agroecological practices must converge to develop crops that can maintain productivity under water-limited conditions. Targeting RSA traits, supported by insights into molecular regulation and environmental responsiveness, will be pivotal in designing resilient agricultural systems for the future.

Acknowledgements

This book chapter was funded by the Research Grant from the University of Sannio, Benevento (Department of Science and Technology).

Conflict of interest

The authors declare no conflict of interest.

Author details

Ilva Licaj, Tahreem Ahmed and Mariapina Rocco*
Department of Science and Technology, University of Sannio, Benevento, Italy

*Address all correspondence to: rocco@unisannio.it

References

[1] Takahashi F, Shinozaki K. Long-distance signaling in plant stress response. Current Opinion in Plant Biology. 2019;**47**:106-111. DOI: 10.1016/j.pbi.2018.10.006

[2] Takahashi F, Kuromori T, Urano K, Yamaguchi-Shinozaki K, Shinozaki K. Drought stress responses and resistance in plants: From cellular responses to long-distance intercellular communication. Frontiers in Plant Science. 2020;**11**:556972. DOI: 10.3389/fpls.2020.556972

[3] Canton H. Food and agriculture Organization of the United Nations—FAO. In: Europa Publications, editor. In: The Europa Directory of International Organizations 2021. 23rd ed. London: Routledge; 2021. pp. 297-305

[4] Sehgal A, Sita K, Siddique KH, Kumar R, Bhogireddy S, Varshney RK, et al. Drought or/and heat-stress effects on seed filling in food crops: Impacts on functional biochemistry, seed yields, and nutritional quality. Frontiers in Plant Science. 2018;**9**:1705. DOI: 10.3389/fpls.2018.01705

[5] Junaid MD, Gokce AF. Global agricultural losses and their causes. Bulletin of Biological and Allied Sciences Research. 2024;**2024**(1):66-66. DOI: 10.54112/bbasr.v2024i1.66

[6] Luo W, Xu C, Ma W, Yue X, Liang X, Zuo X, et al. Effects of extreme drought on plant nutrient uptake and resorption in rhizomatous vs bunchgrass-dominated grasslands. Oecologia. 2018;**188**:633-643. DOI: 10.1007/s00442-018-4232-1

[7] He M, Dijkstra FA. Drought effect on plant nitrogen and phosphorus: A meta-analysis. The New Phytologist. 2014;**204**(4):924-931. DOI: 10.1111/nph.12952

[8] Fahad S, Bajwa AA, Nazir U, Anjum SA, Farooq A, Zohaib A, et al. Crop production under drought and heat stress: Plant responses and management options. Frontiers in Plant Science. 2017;**29**:1147. DOI: 10.3389/fpls.2017.01147

[9] Anjum SA, Xie XY, Wang LC, Saleem MF, Man C, Lei W. Morphological, physiological and biochemical responses of plants to drought stress. African Journal of Agricultural Research. 2011;**6**(9):2026-2032. DOI: 10.5897/AJAR10.027

[10] Zhang H, Zhao Y, Zhu JK. Thriving under stress: How plants balance growth and the stress response. Developmental Cell. 2020;**55**(5):529-543. DOI: 10.1016/j.devcel.2020.10.012

[11] Ogura T, Goeschl C, Filiault D, Mirea M, Slovak R, Wolhrab B, et al. Root system depth in Arabidopsis is shaped by EXOCYST70A3 via the dynamic modulation of auxin transport. Cell. 2019;**178**(2):400-412. DOI: 10.1016/j.cell.2019.06.021

[12] Lynch JP. Steep, cheap and deep: An ideotype to optimize water and N acquisition by maize root systems. Annals of Botany. 2013;**112**(2):347-357. DOI: 10.1093/aob/mcs293

[13] Uga Y, Sugimoto K, Ogawa S, Rane J, Ishitani M, Hara N, et al. Control of root system architecture by DEEPER ROOTING 1 increases rice yield under drought conditions. Nature Genetics. 2013;**45**(9):1097-1102. DOI: 10.1038/ng.2725

[14] Kitomi Y, Kanno N, Kawai S, Mizubayashi T, Fukuoka S, Uga Y. QTLs underlying natural variation of root growth angle among rice cultivars with the same functional allele of DEEPER ROOTING 1. Rice. 2015;**8**:1-2. DOI: 10.1186/s12284-015-0049-2

[15] Segal E, Kushnir T, Mualem Y, Shani U. Water uptake and hydraulics of the root hair rhizosphere. Vadose Zone Journal. 2008;**7**(3):1027-1034. DOI: 10.2136/vzj2007.0122

[16] Ouyang W, Yin X, Yang J, Struik PC. Comparisons with wheat reveal root anatomical and histochemical constraints of rice under water-deficit stress. Plant and Soil. 2020;**452**:547-568. DOI: 10.1007/s11104-020-04581-6

[17] Freschet GT, Roumet C, Comas LH, Weemstra M, Bengough AG, Rewald B, et al. Root traits as drivers of plant and ecosystem functioning: Current understanding, pitfalls and future research needs. The New Phytologist. 2021;**232**(3):1123-1158. DOI: 10.1111/nph.17072

[18] Gao R, Yang X, Liu G, Huang Z, Walck JL. Effects of rainfall pattern on the growth and fecundity of a dominant dune annual in a semi-arid ecosystem. Plant and Soil. 2015;**389**(1):335-347. DOI: 10.1007/s11104-014-2366-4

[19] Suseela V, Tharayil N, Orr G, Hu D. Chemical plasticity in the fine root construct of *Quercus* spp. varies with root order and drought. The New Phytologist. 2020;**228**(6):1835-1851. DOI: 10.1111/nph.16841

[20] Cairns JE, Audebert A, Townend J, Price AH, Mullins CE. Effect of soil mechanical impedance on root growth of two rice varieties under field drought stress. Plant and Soil. 2004;**267**:309-318. DOI: 10.1007/s11104-005-0134-1

[21] Palta JA, Turner NC. Crop root system traits cannot be seen as a silver bullet delivering drought resistance. Plant and Soil. 2019;**439**(1-2):31-43. DOI: 10.1007/s11104-018-3864-6

[22] Mahmood T, Iqbal MS, Li H, Nazir MF, Khalid S, Sarfraz Z, et al. Differential seedling growth and tolerance indices reflect drought tolerance in cotton. BMC Plant Biology. 2022;**22**(1):1-11. DOI: 10.1186/s12870-02203724-4

[23] Padilla FM, Pugnaire FI. Rooting depth and soil moisture control Mediterranean woody seedling survival during drought. Functional Ecology. 2007;**21**(3):489-495. DOI: 10.1016/j.bbamem.2006.02.017

[24] Seneviratne SI, Corti T, Davin EL, Hirschi M, Jaeger EB, Lehner I, et al. Investigating soil moisture–climate interactions in a changing climate: A review. Earth-Science Reviews. 2010;**99**(3-4):125-161. DOI: 10.1016/j.earscirev.2010.02.004

[25] Menge DM, Kameoka E, Kano-Nakata M, Yamauchi A, Asanuma S, Asai H, et al. Drought-induced root plasticity of two upland NERICA varieties under conditions with contrasting soil depth characteristics. Plant Production Science. 2016;**19**(3):389-400. DOI: 10.1080/1343943X.2016.1146908

[26] Lucob-Agustin N, Kawai T, Kano-Nakata M, Suralta RR, Niones JM, Hasegawa T, et al. Morpho-physiological and molecular mechanisms of phenotypic root plasticity for rice adaptation to water stress conditions. Breeding Science. 2021;**71**(1):20-29. DOI: 10.1270/jsbbs.20106

[27] Lombardi E, Ferrio JP, Rodríguez-Robles U, Resco de Dios V, Voltas J.

Ground-penetrating radar as phenotyping tool for characterizing intraspecific variability in root traits of a widespread conifer. Plant and Soil. 2021;**468**(1):319-336. DOI: 10.1007/s11104-021-05135-0

[28] Hartnett DC, Wilson GT, Ott JP, Setshogo M. Variation in root system traits among African semi-arid savanna grasses: Implications for drought tolerance. Austral Ecology. 2013;**38**(4):383-392. DOI: 10.1111/j.1442-9993.2012.02422.x

[29] Fry EL, Evans AL, Sturrock CJ, Bullock JM, Bardgett RD. Root architecture governs plasticity in response to drought. Plant and Soil. 2018;**433**(1-2):189-200. DOI: 10.1007/s11104-018-3824-1

[30] Ali ML, Luetchens J, Nascimento J, Shaver TM, Kruger GR, Lorenz AJ. Genetic variation in seminal and nodal root angle and their association with grain yield of maize under water-stressed field conditions. Plant and Soil. 2015;**397**(1-2):213-225. DOI: 10.1007/s11104-015-2554-x

[31] Lynch JP. Harnessing root architecture to address global challenges. The Plant Journal. 2022;**109**(2):415-431. DOI: 10.1111/tpj.15560

[32] Whitmore AP, Whalley WR. Physical effects of soil drying on roots and crop growth. Journal of Experimental Botany. 2009;**60**(10):2845-2857. DOI: 10.1093/jxb/erp200

[33] Ovalle JF, Arellano EC, Ginocchio R. Trade-offs between drought survival and rooting strategy of two South American Mediterranean tree species: Implications for dryland forests restoration. Forests. 2015;**6**(10):3733-3747. DOI: 10.3390/f6103733

[34] Palta JA, Chen X, Milroy SP, Rebetzke GJ, Dreccer MF, Watt M. Large root systems: Are they useful in adapting wheat to dry environments? Functional Plant Biology. 2011;**38**(5):347-354. DOI: 10.1071/FP11031

[35] Henry A, Cal AJ, Batoto TC, Torres RO, Serraj R. Root attributes affecting water uptake of rice (*Oryza sativa*) under drought. Journal of Experimental Botany. 2012;**63**(13):4751-4763. DOI: 10.1093/jxb/ers150

[36] Song SY, Chen Y, Chen J, Dai XY, Zhang WH. Physiological mechanisms underlying OsNAC5-dependent tolerance of rice plants to abiotic stress. Planta. 2011;**234**:331-345. DOI: 10.1007/s00425-011-1403-2

[37] Hasanuzzaman M, Shabala L, Brodribb TJ, Zhou M, Shabala S. Understanding physiological and morphological traits contributing to drought tolerance in barley. Journal of Agronomy and Crop Science. 2019;**205**(2):129-140. DOI: 10.1111/jac.12307

[38] Robin AHK, Ghosh S, Abu Shahed M. PEG-induced osmotic stress alters root morphology and root hair traits in wheat genotypes. Plants (Basel). 2021;**10**(6):1042. DOI: 10.3390/plants10061042

[39] Xiao S, Liu L, Zhang Y, Sun H, Zhang K, Bai Z, et al. Fine root and root hair morphology of cotton under drought stress revealed with RhizoPot. Journal of Agronomy and Crop Science. 2020;**206**(6):679-693. DOI: 10.1111/jac.12429

[40] Zhang Q, Yuan W, Wang Q, Cao Y, Xu F, Dodd IC, et al. ABA regulation of root growth during soil drying and recovery can involve auxin response. Plant, Cell & Environment.

2022;**45**(3):871-883. DOI: 10.1111/pce.14137

[41] Karanja JK, Aslam MM, Qian Z, Yankey R, Dodd IC, Weifeng X. Abscisic acid mediates drought-enhanced rhizosheath formation in tomato. Frontiers in Plant Science. 2021;**12**:658787. DOI: 10.3389/fpls.2021.658787

[42] Dodd IC. Soil moisture heterogeneity during deficit irrigation alters root-to-shoot signalling of abscisic acid. Functional Plant Biology. 2007;**34**(5):439-448. DOI: 10.1071/FP07009

[43] Pospíšilová J. Interaction of cytokinins and abscisic acid during regulation of stomatal opening in bean leaves. Photosynthetica. 2003;**41**:49-56. DOI: 10.1023/A:1025852210937

[44] Krasensky J, Jonak C. Drought, salt, and temperature stress-induced metabolic rearrangements and regulatory networks. Journal of Experimental Botany. 2012;**63**(4):1593-1608. DOI: 10.1093/jxb/err460

[45] Hu L et al. Comparative analysis of root transcriptome profiles between drought-tolerant and susceptible wheat genotypes in response to water stress. Plant Science. 2018;**272**:276-293. DOI: 10.1016/j.plantsci.2018.03.036

[46] Li H, Mo Y, Cui Q, Yang X, Guo Y, Wei C, et al. Transcriptomic and physiological analyses reveal drought adaptation strategies in drought-tolerant and -susceptible watermelon genotypes. Plant Science. 2019;**278**:32-43. DOI: 10.1016/j.plantsci.2018.10.016

[47] Yang Y, Guo Y, Zhong J, Zhang T, Li D, Ba T, et al. Root physiological traits and transcriptome analyses reveal that root zone water retention confers drought tolerance to Opisthopappus taihangensis. Scientific Reports. 2020;**10**(1):2627. DOI: 10.1038/s41598-020-59399-0

[48] Hura T, Hura K, Dziurka K, Ostrowska A, Bączek-Kwinta R, Grzesiak M. An increase in the content of cell wall-bound phenolics correlates with the productivity of triticale under soil drought. Journal of Plant Physiology. 2012;**169**(17):1728-1736. DOI: 10.1016/j.jplph.2012.07.012

[49] Rehman T, Tabassum B, Yousaf S, Sarwar G, Qaisar U. Consequences of drought stress encountered during seedling stage on physiology and yield of cultivated cotton. Frontiers in Plant Science. 2022;**13**:906444. DOI: 10.3389/fpls.2022.906444

[50] Dalal M, Sahu S, Tiwari S, Rao AR, Gaikwad K. Transcriptome analysis reveals interplay between hormones, ROS metabolism and cell wall biosynthesis for drought-induced root growth in wheat. Plant Physiology and Biochemistry. 2018;**130**:482-492. DOI: 10.1016/j.plaphy.2018.07.035

[51] Karlova R, Boer D, Hayes S, Testerink C. Root plasticity under abiotic stress. Plant Physiology. 2021;**187**(3):1057-1070. DOI: 10.1093/plphys/kiab392

[52] Rowe JH, Topping JF, Liu J, Lindsey K. Abscisic acid regulates root growth under osmotic stress conditions via an interacting hormonal network with cytokinin, ethylene and auxin. The New Phytologist. 2016;**211**(1):225-239. DOI: 10.1111/nph.13882

[53] Manzi M, Pitarch-Bielsa M, Arbona V, Gómez-Cadenas A. Leaf dehydration is needed to induce abscisic acid accumulation in roots of citrus plants. Environmental and Experimental

Botany. 2017;**139**:116-126. DOI: 10.1016/j.envexpbot.2017.05.004

[54] Pierik R, Testerink C. The art of being flexible: How to escape from shade, salt, and drought. Plant Physiology. 2014;**166**(1):5-22. DOI: 10.1104/pp.114.239160

[55] Dietrich D, Pang L, Kobayashi A, Fozard JA, Boudolf V, Bhosale R, et al. Root hydrotropism is controlled via a cortex-specific growth mechanism. Nature Plants. 2017;**3**(6):17057. DOI: 10.1038/nplants.2017.57

[56] Tanaka-Takada N, Kobayashi A, Takahashi H, Kamiya T, Kinoshita T, Maeshima M. Plasma membrane-associated Ca^{2+}-binding protein PCaP1 is involved in root hydrotropism of *Arabidopsis thaliana*. Plant & Cell Physiology. 2019;**60**(6):1331-1341. DOI: 10.1093/pcp/pcz042

[57] Choi H, Park HJ, Park JH, Kim S, Im MY, Seo HH, et al. Arabidopsis calcium-dependent protein kinase AtCPK32 interacts with ABF4, a transcriptional regulator of abscisic acid-responsive gene expression, and modulates its activity. Plant Physiology. 2005;**139**(4):1750-1761. DOI: 10.1104/pp.105.069757

[58] Yu XC, Zhu SY, Gao GF, Wang XJ, Zhao R, Zou KQ, et al. Expression of a grape calcium-dependent protein kinase ACPK1 in *Arabidopsis thaliana* promotes plant growth and confers abscisic acid-hypersensitivity in germination, postgermination growth, and stomatal movement. Plant Molecular Biology. 2007;**64**(5):531-538. DOI: 10.1007/s11103-007-9172-9

[59] Ramachandran P, Wang G, Augstein F, de Vries J, Carlsbecker A. Continuous root xylem formation and vascular acclimation to water deficit involves endodermal ABA signalling via miR165. Development. 2018;**145**(3):dev159202. DOI: 10.1242/dev.159202

[60] Orman-Ligeza B, Morris EC, Parizot B, Lavigne T, Babé A, Ligeza A, et al. The xerobranching response represses lateral root formation when roots are not in contact with water. Current Biology. 2018;**28**(19):3165-3173. DOI: 10.1016/j.cub.2018.07.074

[61] Dinneny JR. Developmental responses to water and salinity in root systems. Annual Review of Cell and Developmental Biology. 2019;**35**:239-257. DOI: 10.1146/annurev-cellbio-100617-062949

[62] Hachez C, Zelazny E, Chaumont F. Modulating the expression of aquaporin genes in planta: A key to understand their physiological functions? Biochimica et Biophysica Acta - Biomembranes. 2006;**1758**(8):1142-1156. DOI: 10.1016/j.bbamem.2006.02.017

[63] Peng Y, Lin W, Cai W. Overexpression of a Panax ginseng tonoplast aquaporin alters salt tolerance, drought tolerance and cold acclimation ability in transgenic Arabidopsis plants. Planta. 2007;**226**(3):729-740. DOI: 10.1007/s00425-007-0520-4

[64] Samarah N, Mullen R, Cianzio S. Size distribution and mineral nutrients of soybean seeds in response to drought stress. Journal of Plant Nutrition. 2004;**27**(5):815-835. DOI: 10.1081/PLN-120030673

[65] González E, Gálvez L, Royuela M, Aparicio-Tejo P, Arrese-Igor C. Insights into the regulation of nitrogen fixation in pea nodules: Lessons from drought, abscisic acid and increased photoassimilate availability. Agronomie.

2001;**21**(6-7):607-613. DOI: 10.1051/agro:2001151

[66] Zare M, Nejad MG, Bazrafshan F. Influence of drought stress on some traits in five mung bean (*Vigna radiata* (L.) R. Wilczek) genotypes. International Journal of Agronomy and Plant Production. 2012;**3**(7):234-240. DOI: 10.5555/20123384637

[67] Khan HR, McDonald GK, Rengel Z. Zn fertilization improves water use efficiency, grain yield and seed Zn content in chickpea. Plant and Soil. 2003;**249**:389-400. DOI: 10.1023/A:1022808323744

[68] Kumari VV, Banerjee P, Verma VC, Sukumaran S, Chandran MA, Gopinath KA, et al. Plant nutrition: An effective way to alleviate abiotic stress in agricultural crops. International Journal of Molecular Sciences. 2022;**23**(15):8519. DOI: 10.3390/ijms23158519

[69] Sun L, Song L, Zhang Y, Zheng Z, Liu D. Arabidopsis PHL2 and PHR1 act redundantly as the key components of the central regulatory system controlling transcriptional responses to phosphate starvation. Plant Physiology. 2016;**170**(1):499-514. DOI: 10.1104/pp.15.01336

[70] Swain R, Sahoo S, Behera M, Rout GR. Instigating prevalent abiotic stress resilience in crop by exogenous application of phytohormones and nutrient. Frontiers in Plant Science. 2023;**14**:1104874. DOI: 10.3389/fpls.2023.1104874

[71] Yadav GS, Devi AG, Das A, Kandpal B, Babu S, Das RC, et al. Foliar application of urea and potassium chloride minimizes terminal moisture stress in lentil (*Lens culinaris* L.) crop. Legume Research. 2021;**44**(6):627-633. DOI: 10.18805/LR-4148

[72] Mahouachi J. Changes in nutrient concentrations and leaf gas exchange parameters in banana plantlets under gradual soil moisture depletion. Scientia Horticulturae. 2009;**120**(4):460-466. DOI: 10.1016/j.scienta.2008.12.002

[73] Venugopalan VK, Nath R, Sengupta K, Nalia A, Banerjee S, Chandran MA, et al. The response of lentil (*Lens culinaris* Medik.) to soil moisture and heat stress under different dates of sowing and foliar application of micronutrients. Frontiers in Plant Science. 2021;**12**:679469. DOI: 10.3389/fpls.2021.679469

[74] Tripathi P, Rabara RC, Langum TJ, Boken AK, Rushton DL, Boomsma DD, et al. The WRKY transcription factor family in *Brachypodium distachyon*. BMC Genomics. 2012;**13**:270. DOI: 10.1186/1471-2164-13-270

[75] Tripathi P, Rabara RC, Rushton PJ. A systems biology perspective on the role of WRKY transcription factors in drought responses in plants. Planta. 2014;**239**(2):255-266. DOI: 10.1007/s00425-013-1985-y

[76] Licaj I, Di Meo MC, Fiorillo A, Samperna S, Marra M, Rocco M. Comparative analysis of the response to polyethylene glycol-simulated drought stress in roots from seedlings of "modern" and "ancient" wheat varieties. Plants. 2023;**12**(3):428. DOI: 10.3390/plants12030428

[77] Li A, Zhu L, Xu W, Liu L, Teng G. Recent advances in methods for in situ root phenotyping. PeerJ. 2022;**10**:e13638. DOI: 10.7717/peerj.13638

[78] Zhu X, Liu S, Meng CH, Qin L, Kong L, Xia G. WRKY transcription factors in wheat and their induction by biotic and abiotic stress. Plant Molecular

Biology Reporter. 2013;**31**(5):1053-1067. DOI: 10.1007/s11105-013-0565-4

[79] Wu H, Ni Z, Yao Y, Guo G, Sun Q. Cloning and expression profiles of 15 genes encoding WRKY transcription factor in wheat (*Triticum aestivem* L.). Progress in Natural Science. 2008;**18**(6):697-705. DOI: 10.1016/j.pnsc.2007.12.006

[80] Waadt R, Seller CA, Hsu PK, Takahashi Y, Munemasa S, Schroeder JI. Plant hormone regulation of abiotic stress responses. Nature Reviews. Molecular Cell Biology. 2022;**23**:680-694. DOI: 10.1038/s41580-022-00479-6

[81] Uga Y, Okuno K, Yano M. Dro1, a major QTL involved in deep rooting of rice under upland field conditions. Journal of Experimental Botany. 2011;**62**(8):2485-2494. DOI: 10.1093/jxb/erq429

[82] Feng X, Jia L, Cai Y, Guan H, Zheng D, Zhang W, et al. ABA-inducible DEEPER ROOTING 1 improves adaptation of maize to water deficiency. Plant Biotechnology Journal. 2022;**20**(10):2077-2088. DOI: 10.1111/pbi.13889

[83] Ashraf A, Rehman OU, Muzammil S, Léon J, Naz AA, Rasool F, et al. Evolution of deeper rooting 1-like homoeologs in wheat entails the C-terminus mutations as well as gain and loss of auxin response elements. PLoS One. 2019;**14**:e0214145. DOI: 10.1371/journal.pone.0214145

[84] Uga Y, Kitomi Y, Yamamoto E, Kanno N, Kawai S, Mizubayashi T, et al. A QTL for root growth angle on rice chromosome 7 is involved in the genetic pathway of deeper rooting 1. Rice. 2015;**8**:8. DOI: 10.1186/s12284-015-0044-7

[85] Uga Y, Yamamoto E, Kanno N, Kawai S, Mizubayashi T, Fukuoka S. A major QTL controlling deep rooting on rice chromosome 4. Scientific Reports. 2013;**3**:3040. DOI: 10.1038/srep03040

[86] Uga Y, Hanzawa E, Nagai S, Sasaki K, Yano M, Sato T. Identification of qSOR1, a major rice QTL involved in soil-surface rooting in paddy fields. Theoretical and Applied Genetics. 2012;**124**:75-86. DOI: 10.1007/s00122-011-1688-3

[87] Kitomi Y, Hanzawa E, Kuya N, Inoue H, Hara N, Kawai S, et al. Root angle modifications by the DRO1 homolog improve rice yields in saline paddy fields. Proceedings of the National Academy of Sciences of the United States of America. 2020;**117**(35):21242-21250. DOI: 10.1073/pnas.2005911117

[88] Schneider HM, Lor VSN, Hanlon MT, Perkins A, Kaeppler SM, Borkar AN, et al. Root angle in maize influences nitrogen capture and is regulated by calcineurin B-like protein (CBL)-interacting serine/threonine-protein kinase 15 (ZmCIPK15). Plant, Cell & Environment. 2022;**45**(3):837-853. DOI: 10.1111/pce.14135

[89] Li C, Guo J, Wang D, Chen X, Guan H, Li Y, et al. Genomic insight into changes of root architecture under drought stress in maize. Plant, Cell & Environment. 2023;**46**:1860-1872. DOI: 10.1111/pce.14567

[90] Ren W, Zhao L, Liang J, Wang L, Chen L, Li P, et al. Genome-wide dissection of changes in maize root system architecture during modern breeding. Nature Plants. 2022;**8**:1408-1422. DOI: 10.1038/s41477-022-01274-z

[91] Ma H, Liu C, Li Z, Ran Q, Xie G, Wang B, et al. ZmbZIP4 contributes to stress resistance in maize by regulating ABA synthesis and root development.

Plant Physiology. 2018;**178**(2):753-770. DOI: 10.1104/pp.18.00436

[92] Gao J, Zhao Y, Zhao Z, Liu W, Jiang C, Li J, et al. RRS1 shapes robust root system to enhance drought resistance in rice. The New Phytologist. 2023;**238**(3):1146-1162. DOI: 10.1111/nph.18775

[93] Jung H, Chung PJ, Park SH, Redillas M, Kim YS, Suh JW, et al. Overexpression of OsERF48 causes regulation of OsCML16, a calmodulin-like protein gene that enhances root growth and drought tolerance. Plant Biotechnology Journal. 2017;**15**(10):1295-1308. DOI: 10.1111/pbi.12716

[94] Okushima Y, Fukaki H, Onoda M, Theologis A, Tasaka M. ARF7 and ARF19 regulate lateral root formation via direct activation of LBD/ASL genes in *Arabidopsis*. The Plant Cell. 2007;**19**(1):118-130. DOI: 10.1105/tpc.106.047761

[95] Orosa-Puente B, Leftley N, von Wangenheim D, Banda J, Srivastava AK, Hill K, et al. Root branching toward water involves posttranslational modification of transcription factor ARF7. Science. 2018;**362**(6421):1407-1410. DOI: 10.1126/science.aau3956

[96] Geng D, Chen P, Shen X, Zhang Y, Li X, Jiang L, et al. MdMYB88 and MdMYB124 enhance drought tolerance by modulating root vessels and cell walls in apple. Plant Physiology. 2018;**178**(3):1296-1309. DOI: 10.1104/pp.18.00502

[97] Tan W, Zhang D, Zhou H, Zheng T, Yin Y, Lin H. Transcription factor HAT1 is a substrate of SnRK2.3 kinase and negatively regulates ABA synthesis and signaling in Arabidopsis responding to drought. PLoS Genetics. 2018;**14**(4):e1007336. DOI: 10.1371/journal.pgen.1007336

[98] Chaves-Sanjuan A, Gnesutta N, Gobbini A, Martignago D, Bernardini A, Fornara F, et al. Structural determinants for NF-Y subunit organization and NF-Y/DNA association in plants. The Plant Journal. 2021;**105**(1):49-61. DOI: 10.1111/tpj.15038

[99] Zhang Y, Du H, Gui Y, Xu F, Liu J, Zhang J, et al. Moderate water stress in rice induces rhizosheath formation associated with abscisic acid and auxin responses. Journal of Experimental Botany. 2020;**71**(9):2740-2751. DOI: 10.1093/jxb/eraa021

[100] Wang X, Wang H, Liu S, Ferjani A, Li J, Yan J, et al. Genetic variation in ZmVPP1 contributes to drought tolerance in maize seedlings. Nature Genetics. 2016;**48**:1233-1241. DOI: 10.1038/ng.3636

[101] Zhang X, Mi Y, Mao H, Liu S, Chen L, Qin F. Genetic variation in ZmTIP1 contributes to root hair elongation and drought tolerance in maize. Plant Biotechnology Journal. 2020;**18**:1271-1283. DOI: 10.1111/pbi.13290

[102] Zhang Y, Wang X, Luo Y, Zhang L, Yao Y, Han L, et al. OsABA8ox2, an ABA catabolic gene, suppresses root elongation of rice seedlings and contributes to drought response. Crop Journal. 2020;**8**:480-491. DOI: 10.1016/j.cj.2019.08.006

[103] Werner T, Nehnevajova E, Köllmer I, Novák O, Strnad M, Krämer U, et al. Root-specific reduction of cytokinin causes enhanced root growth, drought tolerance, and leaf mineral enrichment in *Arabidopsis* and tobacco. The Plant Cell. 2010;**22**:3905-3920. DOI: 10.1105/tpc.109.072694

[104] Placido F, Sandhu J, Sato SJ, Nersesian N, Quach T, Clemente TE, et al. The LATERAL ROOT DENSITY gene regulates root growth during water stress in wheat. Plant Biotechnology Journal. 2020;**18**:1955-1968. DOI: 10.1111/pbi.13355

[105] Valifard M, Le Hir R, Müller J, Scheuring D, Neuhaus HE, Pommerrenig B. Vacuolar fructose transporter SWEET17 is critical for root development and drought tolerance. Plant Physiology. 2021;**187**:2716-2730. DOI: 10.1093/plphys/kiab436

[106] Chen Q, Hu T, Li X, Song CP, Zhu JK, Chen L, et al. Phosphorylation of SWEET sucrose transporters regulates plant root: Shoot ratio under drought. Nature Plants. 2022;**8**:68-77. DOI: 10.1038/s41477-021-01040-7

[107] Redillas MC, Jeong JS, Kim YS, Jung H, Bang SW, Choi YD, et al. The overexpression of OsNAC9 alters the root architecture of rice plants enhancing drought resistance and grain yield under field conditions. Plant Biotechnology Journal. 2012;**10**:792-805. DOI: 10.1111/j.1467-7652.2012.00697.x

[108] Jeong JS, Kim YS, Redillas MC, Jang G, Jung H, Bang SW, et al. OsNAC5 overexpression enlarges root diameter in rice plants leading to enhanced drought tolerance and increased grain yield in the field. Plant Biotechnology Journal. 2013;**11**:101-114. DOI: 10.1111/pbi.12011

[109] Lee DK, Jung H, Jang G, Jeong JS, Kim YS, Ha SH, et al. Overexpression of the OsERF71 transcription factor alters rice root structure and drought resistance. Plant Physiology. 2016;**172**:575-588. DOI: 10.1104/pp.16.00379

[110] Tang N, Shahzad Z, Lonjon F, Loudet O, Vailleau F, Maurel C. Natural variation at XND1 impacts root hydraulics and trade-off for stress responses in *Arabidopsis*. Nature Communications. 2018;**9**:3884. DOI: 10.1038/s41467-018-06430-8

[111] Li C, Li L, Reynolds MP, Wang J, Chang X, Mao X, et al. Recognizing the hidden half in wheat: Root system attributes associated with drought tolerance. Journal of Experimental Botany. 2021;**72**(14):5117-5133. DOI: 10.1093/jxb/erab124

[112] Kim Y, Chung YS, Lee E, Tripathi P, Heo S, Kim KH. Root response to drought stress in rice (*Oryza sativa* L.). International Journal of Molecular Sciences. 2020;**21**(4):1513. DOI: 10.3390/ijms21041513

[113] Siddiqui MN, Leon J, Naz AA, Ballvora A. Genetics and genomics of root system variation in adaptation to drought stress in cereal crops. Journal of Experimental Botany. 2021;**72**(4):1007-1019. DOI: 10.1093/jxb/eraa487

[114] Janiak A, Kwasniewski M, Szarejko I. Gene expression regulation in roots under drought. Journal of Experimental Botany. 2016;**67**(4):1003-1014. DOI: 10.1093/jxb/erv512

[115] Ranjan A, Sinha R, Singla-Pareek SL, Pareek A, Singh AK. Shaping the root system architecture in plants for adaptation to drought stress. Physiologia Plantarum. 2022;**174**(2):e13651. DOI: 10.1111/ppl.13651

[116] Ayalew H, Liu H, Yan G. Identification and validation of root length QTLs for water stress resistance in hexaploid wheat (*Triticum aestivum* L.). Euphytica. 2017;**213**:126. DOI: 10.1007/s10681-017-1914-4

[117] Li T, Yang H, Zhang W, Xu D, Dong Q, Wang F, et al. Comparative

transcriptome analysis of root hairs proliferation induced by water deficiency in maize. Journal of Plant Biology. 2017;**60**(1):26-34. DOI: 10.1007/s12374-016-0412

[118] Abdirad S, Ghaffari MR, Majd A, Irian S, Soleymaniniya A, Daryani P, et al. Genome-wide expression analysis of root tips in contrasting rice genotypes revealed novel candidate genes for water stress adaptation. Frontiers in Plant Science. 2022;**13**:792079. DOI: 10.3389/fpls.2022.792079

[119] Wang LQ, Li Z, Wen SS, Wang JN, Zhao ST. WUSCHEL-related homeobox gene PagWOX11/12a responds to drought stress by enhancing root elongation and biomass growth in poplar. Journal of Experimental Botany. 2020;**71**(4):1503-1513. DOI: 10.1093/jxb/erz490

[120] Dash M, Yordanov YS, Georgieva T, Wei H, Busov V. Gene network analysis of poplar root transcriptome in response to drought stress identifies a PtaJAZ3-PtaRAP2.6-centered hierarchical network. PLoS One. 2018;**13**(12):e0208560. DOI: 10.1371/journal.pone.0208560

[121] Dash M, Yordanov YS, Georgieva T, Tschaplinski TJ, Yordanova E, Busov V. Poplar PtabZIP1-like enhances lateral root formation and biomass growth under drought stress. The Plant Journal. 2017;**89**(4):692-705. DOI: 10.1111/tpj.13413

[122] Araus JL, Kefauver SC. Breeding to adapt agriculture to climate change: Affordable phenotyping solutions. Current Opinion in Plant Biology. 2018 Oct;**1**(45):237-247. DOI: 10.1016/j.pbi.2018.05.003

[123] Tracy SR, Nagel KA, Postma JA, Fassbender H, Wasson A, Watt M. Crop improvement from phenotyping roots: Highlights reveal expanding opportunities. Trends in Plant Science. 2020;**25**(1):105-118. DOI: 10.1016/j.tplants.2019.10.015

[124] Fiorani F, Schurr U. Future scenarios for plant phenotyping. Annual Review of Plant Biology. 2013;**64**:267-291. DOI: 10.1146/annurev-arplant-050312-120137

[125] Furbank RT, Tester M. Phenomics—Technologies to relieve the phenotyping bottleneck. Trends in Plant Science. 2011;**16**(12):635-644. DOI: 10.1016/j.tplants.2011.09.005

[126] Baykalov P, Bussmann B, Nair R, Smith AG, Bodner G, Hadar O, et al. Semantic segmentation of plant roots from RGB (mini-) rhizotron images—Generalisation potential and false positives of established methods and advanced deep-learning models. Plant Methods. 2023;**19**(1):122. DOI: 10.1186/s13007-023-01101-2

[127] Rahman G, Sohag H, Chowdhury R, Wahid KA, Dinh A, Arcand M, et al. SoilCam: A fully automated minirhizotron using multispectral imaging for root activity monitoring. Sensors. 2020;**20**(3):787. DOI: 10.3390/s20030787

[128] Wasson AP, Richards RA, Chatrath R, Misra SC, Prasad SVS, Rebetzke GJ, et al. Traits and selection strategies to improve root systems and water uptake in water-limited wheat crops. Journal of Experimental Botany. 2012;**63**(9):3485-3498. DOI: 10.1093/jxb/ers111

[129] Metzner R, van Dusschoten D, Bühler J, Schurr U, Jahnke S. Belowground plant development measured with magnetic resonance imaging (MRI): Exploiting the potential for non-invasive trait quantification

using sugar beet as a proxy. Frontiers in Plant Science. 2014;**5**:469. DOI: 10.3389/fpls.2014.00469

[130] Wasaya A, Zhang X, Fang Q, Yan Z. Root phenotyping for drought tolerance: A review. Agronomy. 2018;**8**(11):241. DOI: 10.3390/agronomy8110241

[131] Meixner M, Kochs J, Foerst P, Windt CW. An integrated magnetic resonance plant imager for mobile use in greenhouse and field. Journal of Magnetic Resonance. 2021;**323**:106879. DOI: 10.1016/j.jmr.2020.106879

[132] Van der Bom FJ, Lösel PD, Kingston AM, Weng ZH, Raymond NS, Maksimenko A, et al. Synchrotron-based 3D X-ray computed tomography reveals root system architecture: Plastic responses to phosphorus placement. Geoderma. 2025;**457**:117299. DOI: 10.1016/j.geoderma.2025.117299

[133] Rogers HH, Bottomley PA. In situ nuclear magnetic resonance imaging of roots: Influence of soil type, ferromagnetic particle content, and soil water. Agronomy Journal. 1962;**79**(6):957-965. DOI: 10.2134/agronj1987.00021962007900060003x

[134] Pflugfelder D, Kochs J, Koller R, Jahnke S, Mohl C, Pariyar S, et al. The root system architecture of wheat establishing in soil is associated with varying elongation rates of seminal roots: Quantification using 4D magnetic resonance imaging. Journal of Experimental Botany. 2022;**5**(73):2050-2060. DOI: 10.1093/jxb/erab551

[135] Michels V, Chou C, Weigand M, et al. Quantitative phenotyping of crop roots with spectral electrical impedance tomography: A rhizotron study with optimized measurement design. Plant Methods. 2024;**20**:118. DOI: 10.1186/s13007-024-01247-7

[136] Basso B, Amato M, Bitella G, Rossi R, Kravchenko A, Sartori L, et al. Two-dimensional spatial and temporal variation of soil physical properties in tillage systems using electrical resistivity tomography. Agronomy Journal. 2010;**102**(2):440-449. DOI: 10.2134/agronj2009.0298

[137] Dalton FN. In-situ root extent measurements by electrical capacitance methods. Plant and Soil. 1995;**173**:157-165. DOI: 10.1007/BF00155527

[138] Cseresnyés I, Kabos S, Takács T, Végh KR, Vozáry E, Rajkai K. An improved formula for evaluating electrical capacitance using the dissipation factor. Plant and Soil. 2017;**419**(1):237-256. DOI: 10.1007/s11104-017-3336-4

[139] Cseresnyés I, Kelemen B, Takács T, Füzy A, Kovács R, Megyeri M, et al. Electrical capacitance versus minirhizotron technique: A study of root dynamics in wheat–pea intercrops. Plants. 2021;**10**(10):1991. DOI: 10.3390/plants10101991

[140] Ellis TW, Murray W, Paul K, Kavalieris L, Brophy J, Williams C, et al. Electrical capacitance as a rapid and non-invasive indicator of root length. Tree Physiology. 2013;**33**(1):3-17. DOI: 10.1093/treephys/tps115

[141] Delgado A, Hays DB, Bruton RK, Ceballos H, Novo A, Boi E, et al. Ground penetrating radar: A case study for estimating root bulking rate in cassava (*Manihot esculenta* Crantz). Plant Methods. 2017;**13**(1):65. DOI: 10.1186/s13007-017-0216-0

[142] Loiseau B, Carrière SD, Jougnot D, Singha K, Mary B, Delpierre N, et al. The geophysical toolbox applied to forest ecosystems–A review. Science of the Total Environment.

2023;**899**:165503. DOI: 10.1016/j.scitotenv.2023.165503

[143] Ruiz-Guzman H, Adams T, Agbona A, Wolfe M, Everett M, Chamberland JF, et al. Thresholding and continuous wavelet transform (CWT) analysis of ground penetrating radar (GPR) data for estimation of potato biomass. Computers and Electronics in Agriculture. 2025;**232**:110114. DOI: 10.1016/j.compag.2025.110114

[144] Lopez-Guerrero MG, Wang P, Phares F, Schachtman DP, Alvarez S, van Dijk K. A glass bead semi-hydroponic system for intact maize root exudate analysis and phenotyping. Plant Methods. 2022;**18**(1):25. DOI: 10.1186/s13007-022-00856-4

[145] Zhu J, Kaeppler SM, Lynch JP. Mapping of QTLs for lateral root branching and length in maize (*Zea mays* L.) under differential phosphorus supply. Theoretical and Applied Genetics. 2005;**111**(4):688-695. DOI: 10.1007/s00122-005-2051-3

[146] Sharma S, Carena MJ. BRACE: A method for high throughput maize phenotyping of root traits for short-season drought tolerance. Crop Science. 2016;**56**(6):2996-3004. DOI: 10.2135/cropsci2016.02.0116

[147] Krzyzaniak Y, Cointault F, Loupiac C, Bernaud E, Ott F, Salon C, et al. In situ phenotyping of grapevine root system architecture by 2D or 3D imaging: Advantages and limits of three cultivation methods. Frontiers in Plant Science. 2021;**12**:638688. DOI: 10.3389/fpls.2021.638688

[148] Licaj I, Felice D, Germinario C, Fiorillo A, Marra M, Rocco M. An artificial intelligence-integrated analysis of the effect of drought stress on root traits of "modern" and "ancient" wheat varieties. Frontiers in Plant Science. 2023;**14**:1241281. DOI: 10.3389/fpls.2023.1241281